Lecture Notes in Statistics

Edited by D. Brillinger, S. Fienberg, J. Gani,
J. Hartigan, and K. Krickeberg

11

Des F. Nicholls
Barry G. Quinn

Random Coefficient Autoregressive Models: An Introduction

Springer-Verlag
New York Heidelberg Berlin

Des F. Nicholls
Reader in Statistics
Australian National University
Canberra
Australia

Barry G. Quinn
Lecturer in Statistics
University of Wollongong
Wollongong
Australia

AMS Classification: 62H99, 62J02, 62J99, 62K99, 62L99

Library of Congress Cataloging in Publication Data
Nicholls, Des F.
Random coefficient autoregressive models.
(Lecture notes in statistics; v. 11)
Bibliography: p.
Includes index.
1. Regression analysis. 2. Random variables.
I. Quinn, Barry G. II. Title. III. Series: Lecture notes in statistics (Springer-Verlag); v. 11.
QA278.2.N5 1982 519.5'36 82-10619

With 11 Illustrations

9 8 7 6 5 4 3 2 1

ISBN-13: 978-0-387-90766-6 e-ISBN-13: 978-1-4684-6273-9
DOI: 10.1007/978-1-4684-6273-9

PREFACE

In this monograph we have considered a class of autoregressive models whose coefficients are random. The models have special appeal among the non-linear models so far considered in the statistical literature, in that their analysis is quite tractable. It has been possible to find conditions for stationarity and stability, to derive estimates of the unknown parameters, to establish asymptotic properties of these estimates and to obtain tests of certain hypotheses of interest.

We are grateful to many colleagues in both Departments of Statistics at the Australian National University and in the Department of Mathematics at the University of Wollongong. Their constructive criticism has aided in the presentation of this monograph. We would also like to thank Dr M.A. Ward of the Department of Mathematics, Australian National University whose program produced, after minor modifications, the "three dimensional" graphs of the log-likelihood functions which appear on pages 83-86.

Finally we would like to thank J. Radley, H. Patrikka and D. Hewson for their contributions towards the typing of a difficult manuscript.

CONTENTS

CHAPTER 1

INTRODUCTION

1.1 Introduction

Until recently the models considered for time series have usually been linear with constant coefficients. In most situations one would not expect such models to be the "best" class of model to fit to a set of real data, although one tacitly makes the assumption that the linear model under consideration is a close approximation to physical reality. A number of factors have resulted in a consideration of different classes of non-linear models, not the least of which is that the theory of linear models is essentially complete. A large amount of the research into these models is now being concentrated on the construction and application of computationally efficient algorithms to determine order and obtain estimates of the unknown parameters which have desirable statistical properties. The increased power and speed of modern computers has also had a significant effect on the direction in which time series research has headed. This is clearly demonstrated for example by the computational requirements of Akaike's criterion (see Akaike (1978)) to determine the order of a particular linear time series model. With the increase in computer capabilities the application of such criteria has become routine.

The steadily increasing interest in various classes of non-linear time series models is clearly demonstrated by the time series literature over the past decade. Granger and Andersen (1978) have introduced the now familiar class of bilinear models (see Robinson (1977) and Subba Rao (1981) also) while random coefficient and time varying parameter models have received attention in both the engineering and econometric literature. Indeed the Annals of Economic and Social Measurement has allocated an entire issue (volume 2, number 4, 1973) to the consideration of such models.

Subba Rao (1970) has discussed autoregressive models with time dependent coefficients and has considered their weighted least squares estimation at a particular instant of time. Tong (1978) and Tong and Lim (1980) have considered threshold autoregressive models, which approximate non-linear time series by means of different linear autoregressive models fitted to subsets of the data, and have discussed the estimation and application of these models to various data sets. Ozaki (1980) has investigated the case of an autoregression for which the coefficients are functions of time which decay exponentially, the exponential term having least effect when a past value of the time series is large, and most effect when the value is small (see Ozaki (1980) p.89-90). The models of Tong, Lim and Ozaki were developed to explain the natural phenomenon known as the limit cycle (see Tong and Lim (1980) p.248).

A class of non-linear models which includes the bilinear, threshold autoregressive and exponential autoregressive models as special cases has been discussed by Priestley (1980). He has described a recursive algorithm for the estimation of these 'state-dependent' models and has shown how such models may be used for forecasting.

Jones (1978) has investigated a first order non-linear autoregression where an observation $X(t)$ at time t is the sum of a fixed non-linear function at time $(t-1)$ and a disturbance term i.e.

$$X(t) = f\{X(t-1)\} + \varepsilon(t) ,$$

where $f(\cdot)$ is the fixed function and $\{\varepsilon(t),\ t = 0, \pm 1, \pm 2, \ldots\}$ is a sequence of identically and independently distributed random variables. Jones has presented methods for approximating the stationary distributions of such processes and derived expressions by which moments, joint moments and densities of stationary processes can be obtained. His theoretical results are illustrated by a number of simulations.

As yet there has been little statistical theory (properties of the estimates, central limit theorems, tests of hypotheses etc.) developed for the bilinear, the threshold autoregressive or the exponential damping coefficient autoregressive models. On the other hand a substantial amount of theory has been developed for certain classes of varying parameter models. Pagan (1980) gives an excellent bibliography of recent contributors who have considered problems associated with these models. In the case of varying parameter models there have as yet, however, been few applications of the theory developed to real data. Kendall (1953) was one of the first to attempt an empirical investigation of such models. He considered a number of economic series and fitted second order autoregressions, the coefficients of which were slowly changing through time as the economy changed. In fact he chose his coefficients to follow quadratic trends. It is enlightening to read this early work of Kendall as it illustrates the point made earlier that developments in computer technology have made it possible for researchers to examine problems which, through computational difficulty, could not have been considered a few years ago. The estimation and interpretation of the spectra of these autoregressive models with time trending coefficients have been considered by Granger and Hatanaka (1964, Chapter 9).

As Kendall (1953) has pointed out, when considering the modelling of economic data, it seems reasonable to generalize the constant coefficient model to one where the constants are themselves changing through time as the economy changes. Kendall, Subba Rao and Jones have restricted their attention to non-linear autoregressive models for which the coefficients, while non-linear, are non-random, while Garbade (1977) has considered the estimation of regression models where the coefficients are assumed to follow a simple random walk. Garbade's approach requires the numerical maximization of a concentrated likelihood function.

A natural variation of these models is the random coefficient autoregressive (RCA) models. These models are in fact the class of model with which we shall be concerned in this monograph. There has been some investigation of these and closely related models in the economic literature. Turnovsky (1968) has considered stochastic models where the errors are multiplicative i.e. models of the form $X(t) = (\alpha+u(t))X(t-1)$, where α is a constant and the $u(t)$ are uncorrelated random variables with $E\{u(t)\} = 0$, $E\{u^2(t)\} = \sigma_t^2$.

More recently, Ledolter (1980) has extended Garbade's (1977) procedure to include autoregressive models, while Conlisk (1974), (1976) has derived conditions for the stability of RCA models. Andel (1976) has argued that when modelling time series data in such fields as hydrology, meteorology and biology, the coefficients of the model under consideration arise "as a result of complicated processes and actions which usually have many random features". This has led him to consider scalar RCA models and to derive conditions for their second order stationarity. In what follows, for certain classes of RCA models, we shall develop a rigorous statistical theory along the lines of that which exists for constant coefficient autoregressions.

A p-variate time series $\{X(t)\}$ will be said to follow a random coefficient autoregressive model of order n, i.e. RCA(n), if $X(t)$ satisfies an equation of the form

$$X(t) = \sum_{i=1}^{n} \{\beta_i + B_i(t)\}X(t-i) + \varepsilon(t) . \tag{1.1.1}$$

For this model the following assumptions are made.

(i) $\{\varepsilon(t);\ t = 0, \pm1, \pm2,\ldots\}$ is an independent sequence of p-variate random variables with mean zero and covariance matrix G.

(ii) The $p \times p$ matrices β_i, $i = 1,\ldots,n$ are constants.

(iii) Letting $B(t) = [B_n(t),\ldots,B_1(t)]$, then $\{B(t);\ t = 0, \pm 1, \pm 2,\ldots.\}$ is an independent sequence of $p \times np$ matrices with mean zero and $E[B(t) \otimes B(t)] = C$. $\{B(t)\}$ is also independent of $\{\varepsilon(t)\}$.

From (1.1.1) it can be seen that if the elements of C are small compared with those of the matrices β_i, then realizations of $\{X(t)\}$ would be expected to resemble realizations of constant coefficient autoregressions. If however it were possible for some $B_i(t)$ to have elements which were large compared with β_i, one might expect to see some large values of $X(t)$ over a long realization, especially if several elements of C were relatively large. Such behaviour would generally be associated with non-stationarity, but may only be an indication of the non-linear nature of the RCA model. The phenomenon is well illustrated in figures 1.1-1.4 where, for samples of size two thousand and for various values of β, C and G, a number of scalar RCA(1) models have been simulated.

In chapter 2 we shall derive conditions for the second order stationarity of models of the form (1.1.1) generalizing Andel's (1976) work, which is concerned with a similar problem for scalar RCA models. The latter part of chapter 2 considers conditions for stability and the relationship between stability and stationarity. Chapter 3-4 will be concerned with the estimation (both least squares and maximum likelihood) of scalar RCA models, as well as a derivation of the asymptotic properties of the estimates. Chapter 5 presents the results of a number of computer experiments (using simulated data) which illustrate the theoretical procedures and results developed in the previous two chapters. Chapter 6 examines the problem of testing the randomness of the coefficients of the model (1.1.1), while chapter 7 discusses the estimation of multivariate RCA models. The final chapter considers an application of the theory developed to the well known Canadian lynx series. An RCA(2) model has been fitted to the first 100 observations of this data set and then used to forecast the next 14 observations. The forecasts obtained are compared to those based on a number of linear models which have been fitted to the lynx data.

In appendix 1.1 of this introductory chapter, for completeness we present a number of useful results from matrix theory, particularly with respect to the Kronecker or tensor product of vectors and matrices. Appendix 1.2 contains a statement of a martingale central limit theorem due to Billingsley (1961) which will be used in proving the central limit theorems for both least squares and maximum likelihood estimates of the parameters of the RCA model (1.1.1).

FIGURE 1.1

RCA(1): $\beta = .8$, $C = .25$, $G = 1.0$

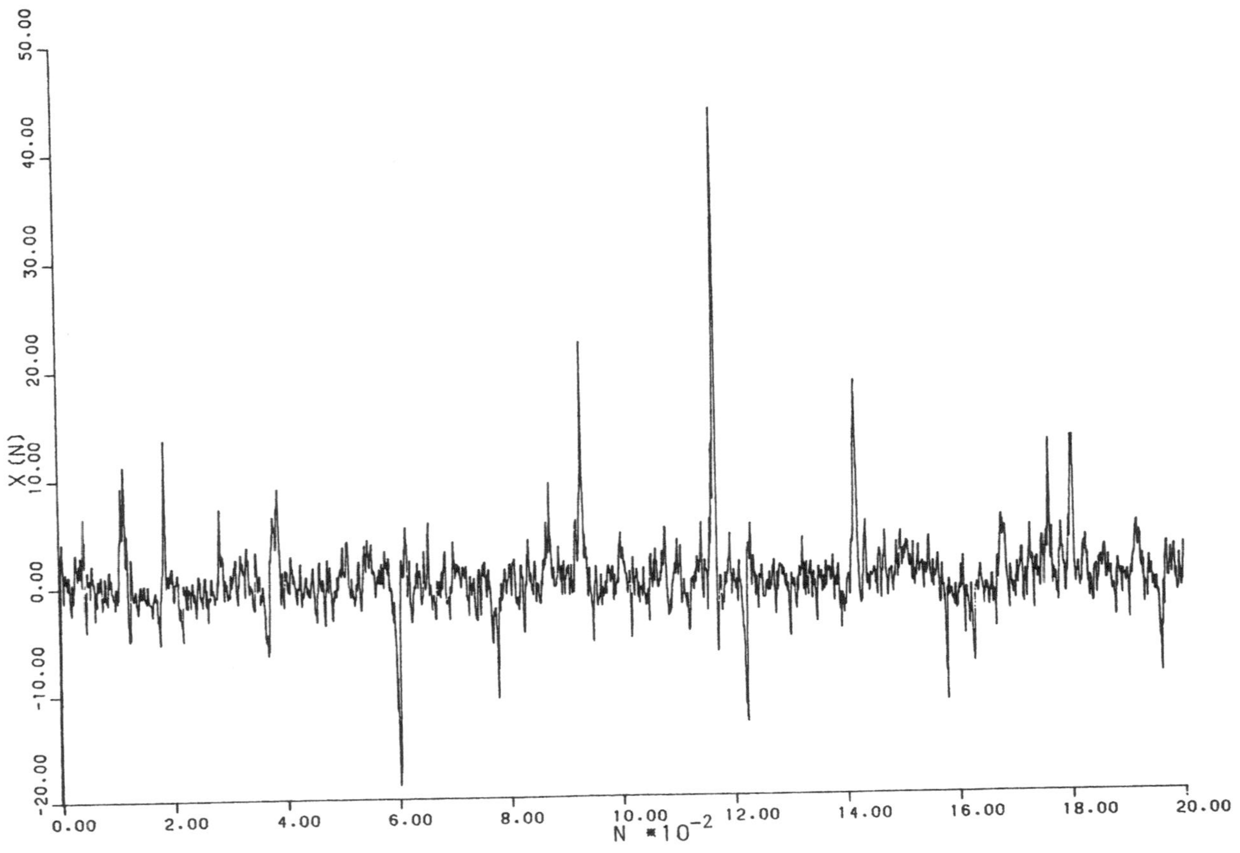

FIGURE 1.2

RCA(1): $\beta = 0$, $C = .81$, $G = 1.0$

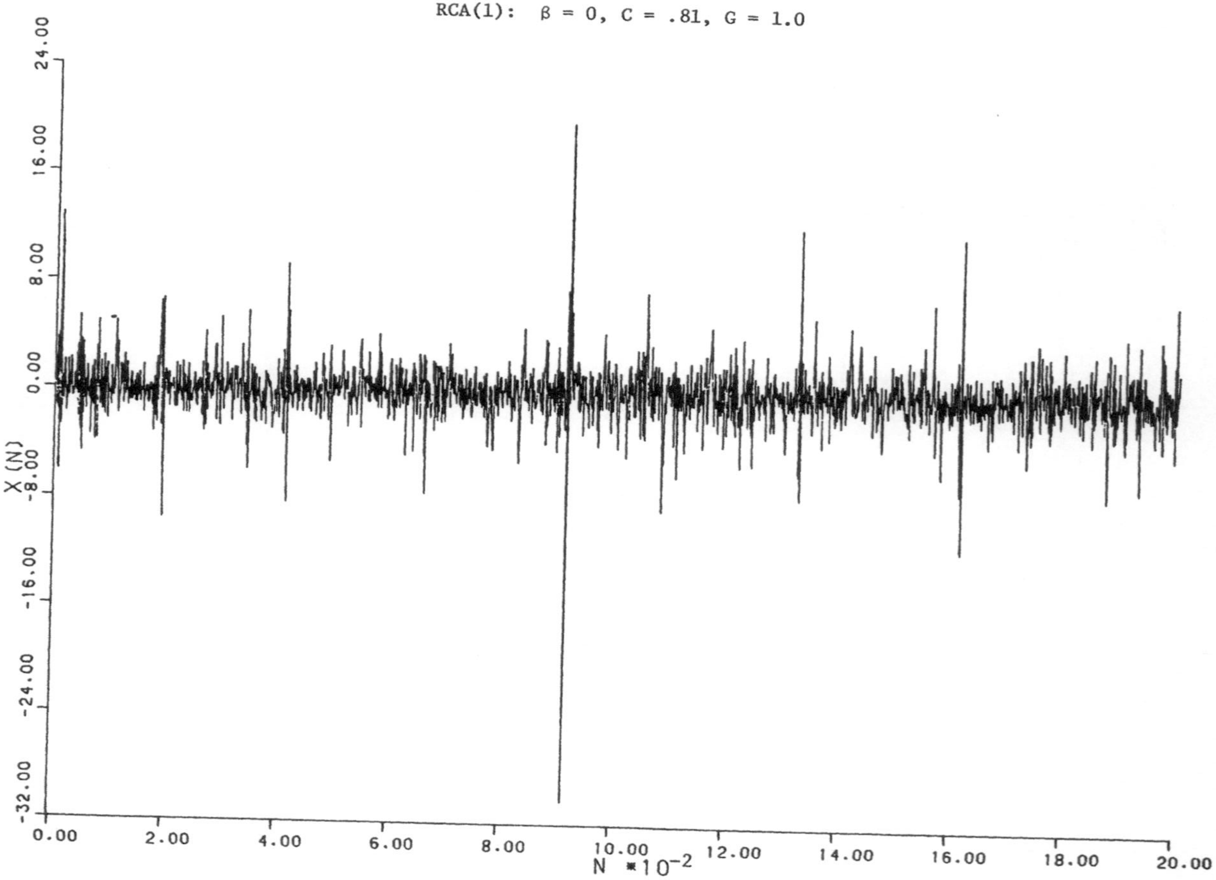

FIGURE 1.3

RCA(1): β = .5, C = .25, G = 1.0

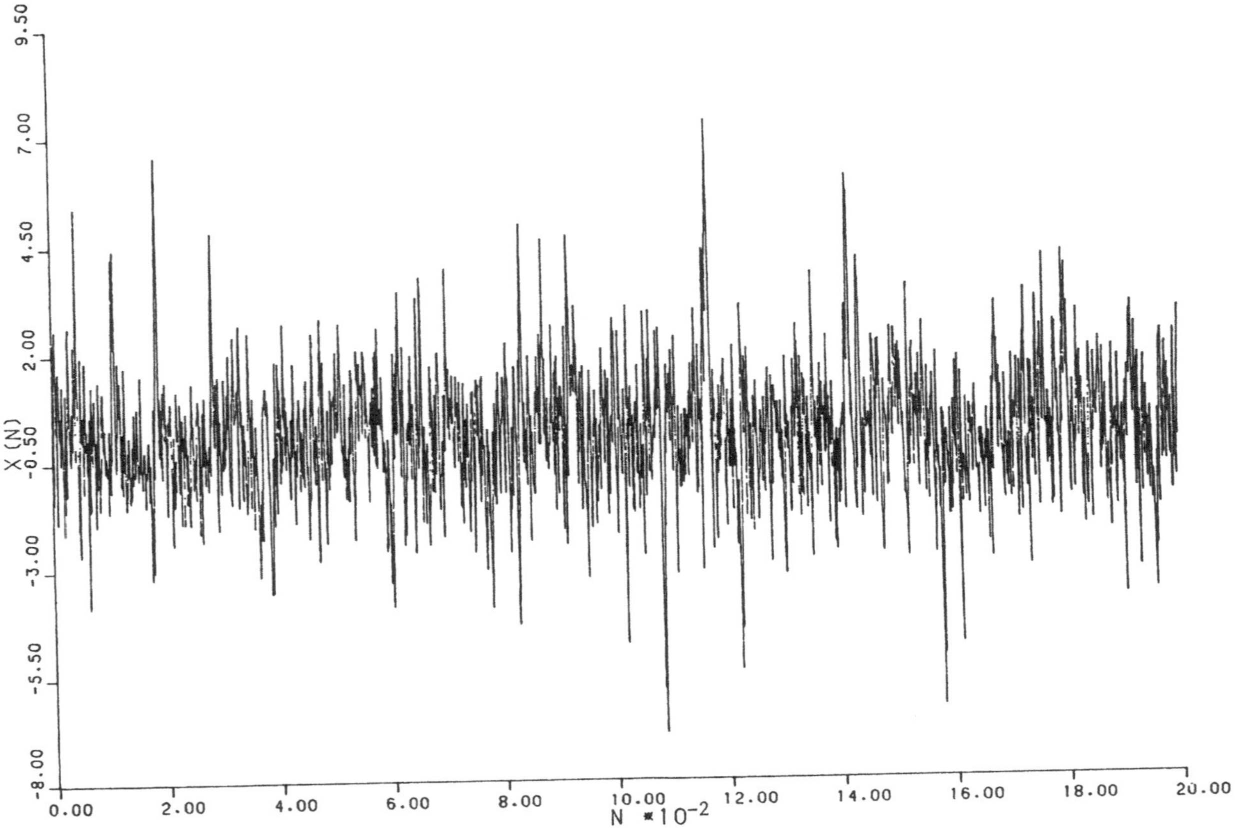

FIGURE 1.4

RCA(1): $\beta = .9$, $C = .16$, $G = 1.0$

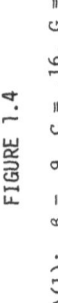

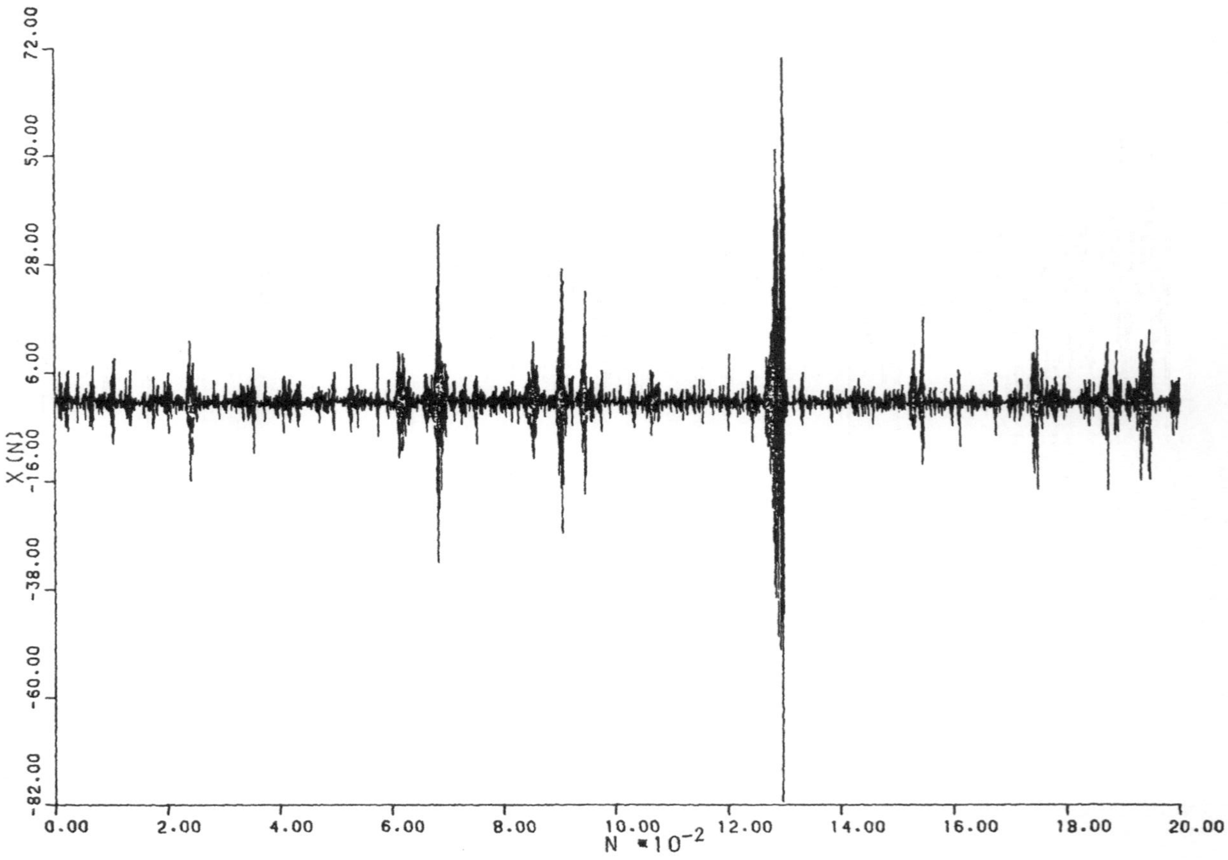

APPENDIX 1.1

The Kronecker (or tensor) notation admits a simple unified approach to the solution of many complicated matrix equations

DEFINITION A.1.1. Let A and B be $m\times n$ and $p\times q$ matrices respectively. Then the Kronecker product $A \otimes B$ of B with A is the $mp\times nq$ matrix whose (i,j)'th block is the $p\times q$ matrix $A_{ij}B$, where A_{ij} is the (i,j)'th element of A.

Next, given any $m\times n$ matrix A, we may define an mn-component vector which has as its elements the elements of A.

DEFINITION A.1.2. Let A be an $m\times n$ matrix. Then the mn-component vector $\text{vec}\, A$ is obtained from A by stacking the columns of A, one on top of the other, in order, from left to right.

The results contained in the following theorem hold for any matrix products which are defined.

THEOREM A.1.1.

1. $\text{vec}(ABC) = (C' \otimes A)\text{vec}\, B$.

2. $\text{tr}(AB) = (\text{vec}(B'))'\text{vec}\, A = (\text{vec}\, B)'\text{vec}(A')$.

3. $(A \otimes B)(C \otimes D) = (AC) \otimes (BD)$.

4. $(A \otimes B)^{-1} = A^{-1} \otimes B^{-1}$, $(A \otimes B)' = A' \otimes B'$.

Proof: See Neudecker (1969).

The results of this theorem will be used repeatedly. In particular, the first result will be used to solve a matrix equation which appears frequently when considering the stationarity of (1.1.1), and which is solved here in more generality.

THEOREM A.1.2. Let V be an $n \times n$ matrix which satisfies the equation

(A.1.1) $$V = MVN' + G$$

where M, N and G are given $n \times n$ matrices. Then if $(I - N \otimes M)$ is invertible, there is a unique solution V which may be obtained from $\text{vec } V = (I - N \otimes M)^{-1} \text{vec } G$.

Proof. Taking the vec of each side of (A.1.1) we obtain

$$\text{vec } V = \text{vec}(MVN') + \text{vec } G = N \otimes M \text{ vec } V + \text{vec } G ,$$

by theorem A.1.1. Thus $(I - N \otimes M)\text{vec } V = \text{vec } G$ and $\text{vec } V = (I - N \otimes M)^{-1}\text{vec } G$. #

If the $n \times n$ matrix A is symmetric, then the $n(n-1)/2$ elements of A above the main diagonal may be obtained from the below-diagonal elements of A. Henderson and Searle (1979) have considered a vector composed of the non-redundant elements of A, which is defined in the following manner.

DEFINITION A.1.3. Let A be an $n \times n$ symmetric matrix. The $n(n+1)/2$-component vector vech A (the "vector-half" of A) is obtained from A by stacking those parts of the columns of A, on and below the main diagonal, one on top of the other in order from left to right.

For symmetric matrices A, it is possible to obtain by linear transformations the vector vec A from the vector vech A, and *vice versa*, which is shown in the following theorem.

THEOREM A.1.3. There exist constant $\{n(n+1)/2\} \times n^2$ matrices K_n and H_n for which $\text{vech } A = H_n \text{vec } A$ and $\text{vec } A = K_n' \text{vech } A$ for any $n \times n$ symmetric matrix A, and $H_n K_n' = I_{n(n+1)/2}$.

Proof. Let H_n be the $\{n(n+1)/2\} \times n^2$ matrix formed by eliminating the $\{(k-1)n+\ell\}$'th rows from I_{n^2} for $2 \le (\ell+1) \le k \le n$. Then it is easy to see that vech A = H_n, since those rows which have been eliminated correspond exactly with the redundant elements of vec A.

The matrix K_n' reinstates the aforementioned redundant elements of vec A, and K_n is constructed by adding the $\{(k-1)n+\ell\}$'th row of I_{n^2} to the $\{(\ell-1)n+k\}$'th row, for $2 \leq (\ell+1) \leq k \leq n$, and then eliminating the former rows.

Now, letting x be any $n(n+1)/2$ component vector, and X the symmetric $n\times n$ matrix for which x = vech X, we have

$$H_nK_n'x = H_n(K_n' \text{vech} X) = H_n \text{vec} X = \text{vech } X = x .$$

Thus $H_nK_n' = I_{n(n+1)/2}$. #

As an example of the above construction we consider the case $n = 2$. The matrices H_2 and K_2 are given by

$$H_2 = \begin{bmatrix} 1 & 0 & 0 & 0 \\ 0 & 1 & 0 & 0 \\ 0 & 0 & 0 & 1 \end{bmatrix}, \quad K_2 = \begin{bmatrix} 1 & 0 & 0 & 0 \\ 0 & 1 & 1 & 0 \\ 0 & 0 & 0 & 1 \end{bmatrix}$$

so that if the 2×2 symmetric matrix

$$A = \begin{bmatrix} a & b \\ b & c \end{bmatrix},$$

then H_2 vec A $= [a,b,c]'$ = vech A ,

while K_2' vech A $= [a,b,b,c]'$ = vec A .

APPENDIX 1.2

THEOREM A.1.4. *Let* $\{\xi_t\}$ *be a sequence of random variables with the property that* ξ_t *may be expressed as a functional, not dependent on* t, *which is measurable with respect to the* σ*-field* F_t *generated by a sequence* $\{\alpha_t, \alpha_{t-1}, \ldots\}$ *of strictly stationary ergodic random variables. Furthermore, suppose that* $E(\xi_t | F_{t-1}) = 0$ *and* $E(\xi_t^2) = c^2 < \infty$. *Then* $(c^2N)^{-\frac{1}{2}} \sum_{t=1}^{N} \xi_t$ *has a distribution which converges to that of a random variable with the standard normal distribution.*

Proof. See Billingsley (1961).

CHAPTER 2

STATIONARITY AND STABILITY

2.1 Introduction

In the case of the scalar RCA model, that is the model with $p = 1$, Andel (1976) has obtained conditions for the existence of a singly infinite process $\{X(t); t = 1-n,\dots,0,1,\dots\}$ satisfying (1.1.1) which is second order stationary. In this chapter we shall extend the results of Andel to the multivariate RCA model and also obtain conditions for the existence of a doubly infinite process $\{X(t); t = 0,\pm 1,\pm 2,\dots\}$ which is second order stationary and satisfies (1.1.1) for all t.

As was mentioned in chapter 1, Conlisk (1974) has found conditions for the (second order) stability of a process $\{X(t)\}$ generated by (1.1.1) for $t \geq 1$, where stability is defined by

DEFINITION 2.1: The process $\{X(t); t = 1,2,\dots\}$ generated by (1.1.1) is said to be stable if, as $t \to \infty$, both

$$E\{X(t)|X(1-n) = x(1-n),\dots,X(0) = x(0)\}$$

and
$$E\{X(t)X'(t-s)|X(1-n) = x(1-n,\dots,X(0) = x(0)\},$$

for fixed $s = 0,1,2,\dots$, converge to finite values not depending on the initial values $\{x(1-n),\dots,x(0)\}$.

The condition for stability was derived by Conlisk (1974). In the latter part of this chapter an alternative condition is derived which is much simpler to check than that of Conlisk. Since we shall in this chapter be concerned mainly with second order effects, a process will be said to be stationary or stable when it is second order stationary or stable respectively, but when required in the strict sense, a stationary process will be said to be strictly stationary.

In deriving necessary and sufficient conditions for stationarity and stability we shall assume conditions (i)-(iii) along with the condition

(iv) there is no non-zero $p\times 1$ constant vector z such that $z'X(t)$ is purely linearly deterministic, that is, is determined exactly as a linear function of $\{X(t-1), X(t-2),\ldots\}$.

Condition (iv) is imposed in order that $X(t)$ have no linearly deterministic component. In fact given (i)-(iii), conditions will be found for (iv) to hold.

2.2 Singly-Infinite Stationarity

DEFINITION 2.2.1. The singly-infinite process $\{X(t);\ t = 0,1,2,\ldots\}$ is stationary if and only if $\mu_i = E(X(i))$, $i \geq 0$, is constant and $V_{ij} = E(X(i)-\mu_i)(X(j)-\mu_j)'$, $i, j \geq 0$ depends only on the value of $(j-i)$.

In order to derive stationarity conditions for processes $\{X(t)\}$ satisfying (1.1.1), it is convenient to represent the nth order RCA model as a higher degree first order model. Define the $np\times 1$ random vector $Y(t)$ by

$$Y(t) = [X'(t+1-n),\ldots,X'(t)]'.$$

Obviously the second order properties of $\{X(t)\}$ may be found from an examination of those of the sequence $\{Y(t)\}$, and vice versa. Equation (1.1.1) may be rewritten in terms of $\{Y(t)\}$ by

$$Y(t) = (M+D(t))Y(t-1) + \eta(t) \tag{2.2.1}$$

where the $np\times np$ matrix M is given by

$$M = \begin{bmatrix} 0 & \vdots & I \\ \cdots & \cdots & \cdots \\ \beta_n & \cdots & \beta_1 \end{bmatrix}$$

The (1,1) block of M is the $(n-1)p\times p$ null matrix, while the (1,2) block is the $(n-1)p\times(n-1)p$ identity. Letting L be the $n\times 1$ vector with only non-zero entry the nth, which is 1, $D(t) = L \otimes B(t)$ and $\eta(t) = L \otimes \varepsilon(t)$. Now

$$E(\eta(t)\eta'(t)) = E[(L \otimes \varepsilon(t))(L \otimes \varepsilon(t))'] = (LL') \otimes E[\varepsilon(t)\varepsilon'(t)] = J \otimes G$$

where $J = LL'$. Letting $\tilde{C} = E[D(t) \otimes D(t)]$, it may be seen that the $\{(n-1)p(np+1+[\frac{k-1}{p}])+k\}$th row of $\tilde{C}$, where $[x]$ denotes the integer part of x, is the kth row of C for $k = 1,\ldots,p^2$, and all the other rows of C are zero.

THEOREM 2.1. <u>The single-infinite process</u> $\{X(t);\ t = 1-n,\ldots,0,1,\ldots\}$ <u>generated by</u> (1.1.1) <u>from</u> $t = 1$ <u>is stationary if and only if</u> $\mu_1 = \mu_0$ <u>and</u> $V_{1,1} = V_{0,0}$, <u>where</u> $\mu_i = E(Y(i))$ <u>and</u> $V_{i,j} = E(Y(i)Y'(j))$.

<u>Proof.</u> The necessity of the conditions is obvious from the construction of $Y(t)$. To prove their sufficiency, we use induction. Assume that $Y(0)$ is independent of $\{\varepsilon(t);\ t = 1,2,\ldots\}$ and $\{B(t);\ t = 1,2,\ldots\}$, and suppose $\mu_t = \mu$, $t = 0,1,\ldots,h$ and $V_{t,t-s} = V_{t-u,t-s-u} = W_s$, $t = s+1,\ldots,h$; $u = 1,\ldots,t-s$; $s = 0,1,\ldots,h$. In the case $h = 1$, these conditions reduce to $\mu_1 = \mu_0 = \mu$ and $V_{1,1} = V_{0,0} = W_0$.

Using (2.2.1) we have, under the induction hypothesis,

$$\begin{aligned} \mu_{h+1} &= E[Y(h+1)] = E[(M+D(h+1))Y(h)+\eta(h+1)] \\ &= ME[Y(h)] = ME[Y(h-1)] = \mu_h, \end{aligned}$$

since $D(h+1)$ and $\eta(h+1)$ are independent of $\{D(1),\ldots,D(h)\}$, $\{\eta(1),\ldots,\eta(h)\}$ and $Y(0)$. For $1 \leq s \leq h$,

$$V_{h+1,h+1-s} = E[Y(h+1)Y'(h+1-s)] = E[(\{M+D(h+1)\}Y(h)+\eta(h+1))Y'(h+1-s)]$$

$$= ME[Y(h)Y'(h+1-s)] = ME[Y(h-1)Y'(h-s)] = E[Y(h)Y'(h-s)] = V_{h,h-s}$$

while

$$V_{h+1,h+1} = E[Y(h+1)Y'(h+1)]$$

$$= E[(\{M+D(h+1)\}Y(h)+\eta(h+1))(\{M+D(h+1)\}Y(h)+\eta(h+1))'] .$$

Thus, eliminating terms which have zero expectations,

$$\text{vec } V_{h+1,h+1} = \text{vec } E[\{M+D(h+1)\}Y(h)Y'(h)\{M+D(h+1)\}'+\eta(h+1)\eta'(h+1)]$$

$$= E[(\{M+D(h+1)\} \otimes \{M+D(h+1)\})\text{vec}(Y(h)Y'(h))] + \text{vec}(J \otimes G)$$

$$= E[M \otimes M+D(h+1) \otimes D(h+1)]E[\text{vec}(Y(h)Y'(h))] + \text{vec}(J \otimes G)$$

$$= (M \otimes M+\tilde{C})\text{vec } V_{h,h} + \text{vec}(J \otimes G)$$

$$= (M \otimes M+\tilde{C}) \text{ vec } V_{h-1,h-1} + \text{vec}(J \otimes G)$$

$$= \text{vec } V_{h,h} \quad . \qquad \#$$

COROLLARY 2.1.1. $\{X(t);\ t = 1-n,\dots,0,1,\dots\}$ generated by (1.1.1) is stationary if and only if $\mu = E[Y(0)]$ satisfies $M\mu = \mu$ and $V = E[Y(0)Y'(0)]$ satisfies

$$\text{vec } V = (M \otimes M+\tilde{C}) \text{ vec } V + \text{vec}(J \otimes G). \tag{2.2.2}$$

Proof. From the proof of the previous theorem, $\mu_1 = E\{Y(1)\} = M\mu_0$ and $\text{vec } V_{1,1} = \text{vec } E[Y(1)Y'(1)] = (M \otimes M+\tilde{C}) \text{ vec } V_{0,0} + \text{vec}(J \otimes G)$. The result now follows directly from theorem 2.1. #

The solution of equation (2.2.2) is left until a later section where it plays an important part in finding conditions for the existence of doubly-infinite stationary solutions to (1.1.1). It should be noted that a solution V to (2.2.2) must be non-negative definite, being a covariance matrix, and positive definite if the assumption (iv) is to be satisfied. Furthermore, there is one solution to (2.2.2) if $(I-M \otimes M-\tilde{C})$ is invertible, namely vec $V = (I - M \otimes M - \tilde{C})^{-1} \text{vec}(J \otimes G)$, but there may exist a solution even if $(I - M \otimes M - \tilde{C})$ is not invertible. This case is considered later.

2.3 Doubly-Infinite Stationarity

Several complications arise when endeavouring to find doubly-infinite processes $X(t)$ which satisfy (1.1.1) for $t = \ldots,-1,0,1,\ldots$. In particular, a solution is required for which $\{X(t)\}$ may be considered as a sequence of random variables on the same probability space as that generated by $\{\varepsilon(t)\}$ and $\{B(t)\}$. Of more use still is a solution for which $\{\varepsilon(t);\ t > s\}$ and $\{B(t);t > s\}$ are independent of $\{X(t);\ t \leq s\}$.

Let F_t be the σ-field generated by $\{(\varepsilon(s),B(s)),\ s \leq t\}$. The above-mentioned properties hold for an F_t-measurable solution $\{X(t)\}$ to (1.1.1), that is, a solution $\{X(t)\}$ for which $X(t)$ is measurable with respect to F_t and $X(t) = U^t X(0)$ where U is the operator which takes $\varepsilon(t)$ to $\varepsilon(t+1)$ and $B(t)$ to $B(t+1)$.

In an attempt to find an F_t-measurable solution to (1.1.1), it is advantageous to obtain a development for $X(t)$ in terms of measurable functions on F_t by iterating the equation (1.1.1), or its counterpart (2.2.1). Defining the matrix product $\prod_{k=i}^{j} A_k$ by

$$\prod_{k=i}^{j} A_k = \begin{cases} A_i A_{i+1} \cdots A_j, & i \leq j, \\ I & , \ j = i - 1, \end{cases}$$

then, if $S_{t,r} = \prod_{k=o}^{r} \{M + D(t-k)\}$; $R_{t,r} = S_{t,r} Y(t-r-1)$ we have, iterating (2.2.1),

$$\text{(2.3.1)} \qquad Y(t) = \{M+D(t)\}(\{M+D(t-1)\}Y(t-2) + \eta(t-1)) + \eta(t)$$

$$= \eta(t) + \{M+D(t)\}\eta(t-1) + \{M+D(t)\}\{M+D(t-1)\}Y(t-2)$$

$$= \sum_{j=o}^{r} S_{t,j-1}\, \eta(t-j) + R_{t,r}$$

which is obtained by induction on r. Furthermore, if $W_{t,r} = Y(t) - R_{t,r}$, noting that $E[\eta(t-i)\eta'(t-j)] = 0$ if $i \neq j$, then

$$\begin{aligned}
\operatorname{vec} E(W_{t,r} W'_{t,r}) &= \operatorname{vec} E\Big[\sum_{j=0}^{r} S_{t,j-1}\eta(t-j)\Big]\Big[\sum_{i=0}^{r} S_{t,i-1}\eta(t-i)\Big]' \\
&= \operatorname{vec} E\Big[\sum_{j=0}^{r} S_{t,j-1}\eta(t-j)\eta'(t-j)S'_{t,j-1}\Big] \\
&= E \sum_{j=0}^{r} [(S_{t,j-1} \otimes S_{t,j-1})\operatorname{vec} \eta(t-j)\eta'(t-j)] \\
&= E \sum_{j=0}^{r} \Big[\prod_{k=0}^{j-1} (\{M+D(t-k)\}\otimes\{M+D(t-k)\})\operatorname{vec}[\eta(t-j)\eta'(t-j)]\Big] \\
&= \sum_{j=0}^{r} \Big[\prod_{k=0}^{j-1} E(\{M+D(t-k)\}\otimes\{M+D(t-k)\})\operatorname{vec} E[\eta(t-j)\eta'(t-j)]\Big] \\
&= \sum_{j=0}^{r} (M \otimes M + \tilde{C})^j \operatorname{vec}(J \otimes G).
\end{aligned}$$

In the above, result 1 of theorem A.1.1 was used, as well as the fact that $(\prod_{i=0}^{j} A_i) \otimes (\prod_{k=0}^{j} B_k) = \prod_{i=0}^{j} (A_i \otimes B_i)$, whenever the matrix products are defined.

It will be seen in what follows that the stationarity of an F_t-measurable solution $\{X(t)\}$ involves the convergence of $\{W_{t,r}\}$ and $\{R_{t,r}\}$ as r increases, for fixed t. The following lemma will prove useful in establishing this convergence, while lemma 2.2 examines the question of uniqueness of solutions.

LEMMA 2.1. <u>If the sum</u> $\sum_{j=0}^{r} (M \otimes M+\tilde{C})^j \operatorname{vec}(J \otimes G)$ <u>converges as</u> $r \to \infty$, <u>and if</u> H <u>is positive definite, where</u> $\operatorname{vec} H = \operatorname{vec} G + C \sum_{j=0}^{\infty} (M \otimes M+\tilde{C})^j \operatorname{vec}(J \otimes G)$, <u>then the matrix</u> M <u>has all its eigenvalues within the unit circle</u>.

<u>Proof.</u> See appendix 2.1.

LEMMA 2.2. If the matrix $(M \otimes M + \tilde{C})$ does not possess an eigenvalue equal to unity, and an F_t-measurable stationary solution exists to (1.1.1), then this solution is the unique F_t-measurable stationary solution.

Proof. See appendix 2.1.

Conditions are now established for the existence of F_t-measurable stationary solutions to (1.1.1).

THEOREM 2.2. In order that there exist a stationary F_t-measurable solution to (1.1.1) satisfying assumptions (i)-(iv), it is necessary that $\sum_{j=0}^{r} (M \otimes M + \tilde{C})^j \operatorname{vec}(J \otimes G)$ converge as $r \to \infty$, and sufficient that this occur with H positive definite where

$$\operatorname{vec} H = \operatorname{vec} G + C \sum_{j=0}^{\infty} (M \otimes M + \tilde{C})^j \operatorname{vec}(J \otimes G) .$$

When $(M \otimes M + \tilde{C})$ does not have a unit eigenvalue, this latter condition is both necessary and sufficient, and there is a unique stationary solution $\{X(t)\}$ obtained from

$$Y(t) = \eta(t) + \sum_{j=1}^{\infty} \left(\prod_{k=0}^{j-1} \{M + D(t-k)\} \right) \eta(t-j) . \tag{2.3.2}$$

Proof. We first show necessity. Suppose $\{X(t)\}$ satisfies (1.1.1) and is F_t-measurable and stationary. Using (2.3.1) and the notation $W_{t,r}$, $S_{t,r}$ and $R_{t,r}$ adopted there, and letting $V = E(Y(t)Y'(t))$, then

$$\begin{aligned} \operatorname{vec} V &= \operatorname{vec}[E(W_{t,r}+R_{t,r})(W_{t,r}+R_{t,r})'] \\ &= \operatorname{vec}[E(W_{t,r}W'_{t,r})+E(R_{t,r}R'_{t,r})+E(R_{t,r}W'_{t,r})+E(W_{t,r}R'_{t,r})]. \end{aligned}$$

Now,

$$\begin{aligned} \operatorname{vec} E(R_{t,r}R'_{t,r}) &= \operatorname{vec} E[S_{t,r}Y(t-r-1)Y'(t-r-1)S'_{t,r}] \\ &= E[(S_{t,r} \otimes S_{t,r}) \operatorname{vec}(Y(t-r-1)Y'(t-r-1))] \\ &= (M \otimes M + \tilde{C})^{r+1} \operatorname{vec} V \end{aligned}$$

and

$$\text{vec } E(W_{t,r}R'_{t,r}) = \text{vec } E[\sum_{j=0}^{r} S_{t,j-1}\ \eta(t-j)Y'(t-r-1)\ S'_{t,j-1}]$$

$$= 0$$

since $E[\eta(t-j)Y'(t-r-1)] = 0$ for $j = 0,\ldots,r$. Thus

(2.3.3) $$\text{vec } V = \sum_{j=0}^{r} (M \otimes M+\tilde{C})^j \text{vec}(J \otimes G) + (M \otimes M+C)^{r+1} \text{ vec } V, \quad r = 0,1,2,\ldots .$$

Let

$$Q_o = J \otimes G;\ Q_j = MQ_{j-1}M' + E\{D(t)Q_{j-1}D'(t)\},\quad j = 1,2,\ldots;$$

$$R_o = V \text{ and } R_j = MR_{j-1}M' + E\{D(t)R_{j-1}D'(t)\},\quad j = 1,2,\ldots .$$

It is clear that each of Q_j, R_j, $j = 0,1,2,\ldots$ is non-negative definite.
Also

$$V = \sum_{j=0}^{r} Q_j + R_{r+1},\quad r = 0,1,2\ldots .$$

If z is any np×1 fixed vector, then

(2.3.4) $$z'Vz = \sum_{j=0}^{r} z'Q_jz + z'R_{r+1}z.$$

Now $\sum_{j=0}^{r} z'Q_jz$ is nondecreasing in r, while $z'Vz$ and $z'R_{r+1}z$ are non-negative. Since (2.3.4) holds for $r = 0,1,2\ldots$, it follows that $\sum_{j=0}^{r} z'Q_jz$ is bounded above by $z'Vz$ and is therefore convergent for every vector z. Thus $\sum_{j=0}^{r} Q_j$ converges, as $r \to \infty$ to a non-negative matrix and so

$$\sum_{j=0}^{r} (M \otimes M+\tilde{C})^j \text{vec}(J \otimes G)$$

converges as $r \to \infty$, as required.

Suppose now that $\sum_{j=0}^{r} (M \otimes M+\tilde{C})^j \text{vec}(J \otimes G)$ converges ar r increases, and that H, given by $\text{vec } H = \text{vec } G + C \sum_{j=0}^{\infty} (M \otimes M+\tilde{C})^j \text{vec}(J \otimes G)$, is positive definite. It has been shown above that the limit $W(t)$ of $W_{t,r}$ as r increases exists in mean square, and thus in probability. Moreover,

$$\{M+D(t)\}W(t-1) = \sum_{j=0}^{\infty} \left(\prod_{k=-1}^{j-1} \{M+D(t-1-k)\}\right)\eta(t-1-j)$$
$$= \sum_{j=1}^{\infty} S_{t,j-1}\eta(t-j)$$
$$= W(t) - \eta(t) .$$

Hence $\{W(t)\}$ satisfies (2.2.1). $\{W(t)\}$ is obviously F_t-measurable, and is also stationary since $\text{vec}E[W(t)W'(t)] = \sum_{j=0}^{\infty} (M \otimes M+\tilde{C})^j \text{vec}(J \otimes G)$ which is finite, $W(t) = U^t W(0)$ because the functional form taken by $W(t)$ does not depend on t, and $\{D(t)\}$ and $\{\eta(t)\}$ are stationary. Let $W(t) = [w'(t+1-n), w'(t+2-n),\ldots, w'(t)]'$ where each $w(s)$ is a $p\times 1$ random vector and suppose there is a $p\times 1$ vector z such that $z'w(t)$ is perfectly linearly predictable, that is, $z'w(t)$ is completely linearly determined by $\{w(t-1),w(t-2),\ldots\}$. Then

$$z'w(t) = E(z'w(t)|F_{t-1})$$
$$= E\{(z'[\sum_{i=1}^{n} \beta_i w(t-i) + \sum_{i=1}^{n} B_i(t)w(t-i) + \varepsilon(t)])|F_{t-1}\}$$
$$= z' \sum_{i=1}^{n} \beta_i w(t-i),$$

since $B(t)$ and $\varepsilon(t)$ are independent of $\{B(t-1),B(t-2),\ldots\}$ and $\{\varepsilon(t-1),\varepsilon(t-2),\ldots\}$, and $W(t)$ satisfies (2.2.1). Thus

$$z'\{\sum_{i=1}^{n} B_i(t)w(t-i)+\varepsilon(t)\} = 0$$

and, since $\sum_{i=1}^{n} B_i(t)w(t-i) = B(t)W(t-1)$,

$$Ez'(B(t)W(t-1)+\varepsilon(t))(W'(t-1)B'(t)+\varepsilon'(t))z = 0 .$$

That is, $z'Hz = 0$, since

$$\text{vec } E(B(t)W(t-1)W'(t-1)B'(t)) = E(B(t) \otimes B(t))\text{vec } E(W(t-1)W'(t-1))$$
$$= C \sum_{j=0}^{\infty} (M \otimes M+\tilde{C})^j \text{vec}(J \otimes G) = \text{vec } H - \text{vec } G.$$

But H is positive definite, so that $z = 0$, and the conditions are sufficient for $\{w(t)\}$ to be an F_t-measurable stationary solution to (1.1.1) satisfying condition (iv).

Of course, if G is positive definite, then H is of necessity positive definite since (H-G) is non-negative definite, and the necessary condition is also sufficient. If G is not positive definite then the sufficient conditions are also necessary when $(M \otimes M+\tilde{C})$ does not have an eigenvalue equal to unity. For then the solution $\{W(t)\}$ is the unique solution, by lemma 2.2. However, if H is not positive definite there exists a p×1 vector z with $z'Hz = 0$ and $z'(B(t)W(t-1)+\varepsilon(t)) = 0$ almost everywhere, and $z'w(t) = z' \sum_{i=1}^{n} \beta_i w(t-1)$ almost everywhere, which is seen by inverting the previous proof, and condition (iv) does not hold. #

COROLLARY 2.2.1. In order that there exist a unique F_t-measurable stationary solution to (1.1.1), it is sufficient that all the eigenvalues of $(M \otimes M+\tilde{C})$ or $h(M \otimes M+\tilde{C})k'$ be less than unity in modulus, where $h = H_{np}$ and $k = K_{np}$, defined in theorem A.1.3.

Proof. $(M \otimes M+\tilde{C})$ may be represented in Jordan canonical form as

$$M \otimes M + \tilde{C} = P\Lambda P^{-1} \tag{2.3.5}$$

where Λ has the eigenvalues of $(M \otimes M+\tilde{C})$ along its main diagonal, and zeros elsewhere, unless $(M \otimes M+\tilde{C})$ has eigenvalues of multiplicity greater than one, in which case there may be several ones in the first upper diagonal. Now, $(M \otimes M+\tilde{C})^j = P\Lambda^j P^{-1}$ and it is well known that if the diagonal elements of Λ are less than unity in modulus, then Λ^j converges to zero at a geometric rate and $\lim_{r\to\infty} \sum_{j=0}^{r} \Lambda^j = (I-\Lambda)^{-1}$. Furthermore,

$$\begin{aligned}\lim_{r\to\infty} \sum_{j=0}^{r} (M \otimes M+\tilde{C})^j \operatorname{vec}(J \otimes G) &= P(I-\Lambda)^{-1}P^{-1}\operatorname{vec}(J \otimes G) \\ &= (I-P\Lambda P^{-1})^{-1}\operatorname{vec}(J \otimes G) \\ &= (I-M \otimes M-\tilde{C})^{-1}\operatorname{vec}(J \otimes G).\end{aligned}$$

Thus, using lemmas 2.1, 2.2 and theorem 2.2, there is a unique F_t-measurable stationary solution given by (2.3.2) if all the eigenvalues of $(M \otimes M+\tilde{C})$ are less than unity in modulus. Furthermore, noting that $\mathrm{vec}(J \otimes G) = k' \,\mathrm{vech}(J \otimes G)$, it follows by induction on r that

$$h \sum_{j=0}^{r} (M \otimes M+\tilde{C})^j \mathrm{vec}(J \otimes G) = \sum_{j=0}^{r} \{h(M \otimes M+\tilde{C})k'\}^j \mathrm{vech}(J \otimes G) .$$

Hence, if the eigenvalues of $(h(M \otimes M+\tilde{C})k')$ are less than unity in modulus, $h \sum_{j=0}^{r} (M \otimes M+\tilde{C})^j \mathrm{vec}(J \otimes G)$ converges as r increases to $(I-h(M \otimes M+\tilde{C})k')^{-1}\mathrm{vech}(J \otimes G)$ by the above argument. However, since $h \sum_{j=0}^{r} (M \otimes M+\tilde{C})^j \mathrm{vec}(J \otimes G)$ is the vech of a symmetric matrix, its convergence is equivalent to the convergence of $\sum_{j=0}^{r} (M \otimes M+\tilde{C})^j \mathrm{vec}(J \otimes G)$, the vec of the same matrix, and the condition that the eigenvalues of $(h(M \otimes M+\tilde{C})k')$ have moduli less than unity is also sufficient. #

It has been seen that the convergence of $\sum_{j=0}^{r} (M \otimes M+\tilde{C})^j \mathrm{vec}(J \otimes G)$ is a central requirement for the existence of an F_t-measurable stationary solution to (1.1.1). Using (2.3.5) we have

$$\sum_{j=0}^{r} (M \otimes M+\tilde{C})^j \mathrm{vec}(J \otimes G) = P(\sum_{j=0}^{r} \Lambda^j)P^{-1}\mathrm{vec}(J \otimes G) \tag{2.3.6}$$

and

$$h \sum_{j=0}^{r} (M \otimes M+\tilde{C})^j \mathrm{vec}(J \otimes G) = Q(\sum_{j=0}^{r} \Omega^j)Q^{-1}\mathrm{vech}(J \otimes G) \quad , \tag{2.3.7}$$

where $h(M \otimes M+\tilde{C})k'$ is represented in Jordan canonical form as $Q\Omega Q^{-1}$. Even if Λ or Ω have eigenvalues whose moduli are larger than or equal to unity, the right hand sides of (2.3.6) and (2.3.7) will converge if $\mathrm{vec}(J \otimes G)$ or

$\text{vech}(J \otimes G)$ are orthogonal to the rows of P^{-1} or Q^{-1}, respectively, corresponding to those diagonal elements of Λ or Ω, respectively, which are greater than or equal to unity in modulus. However, this is impossible when $p = 1$, as is shown in the following corollary to theorem 2.2.

COROLLARY 2.2.2. When $p = 1$ and $G \neq 0$, in order that there exist a unique F_t-measurable stationary solution to (1.1.1) it is necessary and sufficient that the eigenvalues of $(M \otimes M + \tilde{C})$ have moduli less than unity.

Proof. In view of the above remarks it is necessary only to show that the rows of P^{-1} corresponding to a diagonal element λ of Λ for which $|\lambda| \geq 1$ cannot be orthogonal to $\text{vec}(J \otimes G)$. Let z' be one of these rows of P^{-1} for which z' is a left eigenvector of $(M \otimes M + \tilde{C})$, noting that there is at least one such vector. If $z'\text{vec}(J \otimes G) = 0$, then the last element of z is zero, since the only non-zero element of $\text{vec}(J \otimes G)$ is G, its last element. But $\tilde{C}$ has only one non-zero row, its last, which is C. Hence

$$z'(M \otimes M + \tilde{C}) = z'(M \otimes M) .$$

However, $z'(M \otimes M + \tilde{C}) = \lambda z'$ and so λ is also an eigenvalue of $M \otimes M$. By lemma 2.1, all the eigenvalues of M are less than unity in modulus. Let $P\Lambda P^{-1}$ be a Jordan canonical form of M and for general p let $[\lambda_1, \ldots, \lambda_{np}]' = \text{diag}(\Lambda)$. Now, if λ is an eigenvalue of $M \otimes M$, then $\det[\lambda I - M \otimes M] = 0$. But

$$\det[\lambda I - M \otimes M] = \det[\lambda I - (P\Lambda P^{-1}) \otimes (P\Lambda P^{-1})]$$

$$= \det[\lambda I - (P \otimes P)(\Lambda \otimes \Lambda)(P^{-1} \otimes P^{-1})] = \det[\lambda (P \otimes P)(P \otimes P)^{-1} - (P \otimes P)(\Lambda \otimes \Lambda)(P \otimes P)^{-1}]$$

$$= \det[(P \otimes P)(\lambda I - \Lambda \otimes \Lambda)(P \otimes P)^{-1}] = \det(P \otimes P)\det(\lambda I - \Lambda \otimes \Lambda)\det[(P \otimes P)^{-1}]$$

$$= \det(\lambda I - \Lambda \otimes \Lambda) = \prod_{i,j=1}^{np} (\lambda - \lambda_i \lambda_j)$$

since the matrix $(\lambda I - \Lambda \otimes \Lambda)$ has no non-zero sub-diagonal elements. Thus $\lambda = \lambda_i \lambda_j$ for some i and j, and $|\lambda|^2 = |\lambda_i|^2 |\lambda_j|^2$. Since $|\lambda_i|^2 < 1$ for all i, then $|\lambda|^2 < 1$ and $|\lambda| < 1$. #

From (2.3.3), $V = E[Y(t)Y'(t)]$, where $\{X(t)\}$ is an F_t-measurable stationary solution to (1.1.1) satisfying (iv), satisfies the equation

$$(2.3.8) \qquad \text{vec } V = (M \otimes M) \text{ vec } V + \text{vec}(J \otimes H)$$

where $H = G + E[B(t)VB'(t)]$, and is positive definite as is shown in the proof of theorem 2.2. A minor modification of lemma 2.1 shows also that the matrix M has all its eigenvalues inside the unit circle, for the proof will hold when the matrix W is replaced by any matrix V satisfying the equation vec V = $(M \otimes M + \tilde{C})$ vec V + vec$(J \otimes G)$. The above proof shows that the matrix $M \otimes M$ also has all its eigenvalues within the unit circle, so that $(I - M \otimes M)$ is invertible. In fact, the condition that M have all its eigenvalues within the unit circle is easily seen to be equivalent to the condition that $\det\{I - \sum_{i=1}^{n} \beta_i z^i\}$ have all its zeros outside the unit circle (see Andel (1971)). This is exactly the condition that a stationary F_t-measurable solution exist to (1.1.1) with C = 0, that is, B(t) identically zero, (1.1.1) then being the equation for a linear (fixed coefficient) autoregression. Now (2.3.8) may be solved to obtain

$$\text{vec } V = (I - M \otimes M)^{-1} \text{vec}(J \otimes H)$$

and so V is the same matrix obtained by replacing G by H and B(t) by 0 in (1.1.1) and calculating the covariance matrix $E[Y(t)Y'(t)]$ for the resulting solution $\{X(t)\}$. The columns of $(I-M \otimes M)^{-1}$ corresponding to the zero elements of the vector $\text{vec}(J \otimes H)$ will play no part in deducing the covariance structure of $\{X(t)\}$. With this in mind, we define the matrix A as being the $n^2p^2 \times p^2$ matrix formed from those effective columns of $(I-M \otimes M)^{-1}$, that is, the kth column of A is the $\{(n-1)p(np+1 + [\frac{k-1}{p}])+k\}$th column of $(I-M \otimes M)^{-1}$, $k = 1,\dots,p^2$. Thus $\text{vec}\, V = A\, \text{vec}\, H$. As will be shown, the matrix A plays a dual role in the question of the existence of stationary solutions for random coefficient autoregressive models.

THEOREM 2.3. When $(M \otimes M+\tilde{C})$ does not have a unit eigenvalue, there exists a unique $\mathcal{F}_t$-measurable stationary solution $\{X(t)\}$ to (1.1.1) satisfying (iv) if and only if the matrix V given by

(2.3.9) $$\text{vec}\, V = (1-M \otimes M-\tilde{C})^{-1}\text{vec}(J \otimes G)$$

is positive definite. An equivalent condition is that the eigenvalues of M be less than unity in modulus, together with the condition that the matrix H given by

$$\text{vec}\, H = (I-CA)^{-1}\text{vec}\, G$$

be positive definite. The covariance matrix V of Y(t) is then given by

$$\text{vec}\, V = A\, \text{vec}\, H.$$

Proof. Since the solution $\{X(t)\}$, if it exists, is unique, the covariance matrix V of Y(t) is obtained from

$$\text{vec}\, V = \sum_{j=0}^{\infty} (M \otimes M+\tilde{C})^j\text{vec}(J \otimes G)$$

the existence of the solution depending on the convergence of the above sum. Assuming that the sum does converge, i.e. the solution exists, it has already been seen that

$$\text{vec}\, V = (I-M \otimes M-\tilde{C})^{-1}\text{vec}(J \otimes G),$$

$(I-M \otimes M-\tilde{C})$ being invertible since it has no zero eigenvalues. That V being positive definite is both necessary and sufficient is now evident, since $(I-M \otimes M-\tilde{C})^{-1}\text{vec}(J \otimes G)$ is equal to $\sum_{j=0}^{\infty} (M \otimes M+\tilde{C})^j \text{vec}(J \otimes G)$ whenever the latter sum exists by the proof above and by theorem 2.2.

Now, if V is positive definite, then so is the matrix H defined in (2.3.8) and M has its eigenvalues within the unit circle by lemma 2.1. Conversely, if H is positive definite, and M has all its eigenvalues within the unit circle, then V is positive definite also as is seen by employing the following argument used by Andel (1971): since M and $(M \otimes M)$ have their eigenvalues within the unit circle,

$$\text{vec } V = (I-M \otimes M)^{-1}\text{vec}(J \otimes H) = \sum_{j=0}^{\infty} (M \otimes M)^j \text{vec}(J \otimes H)$$

$$= \sum_{j=0}^{\infty} (M^j \otimes M^j)\text{vec}(J \otimes H) = \text{vec}\{ \sum_{j=0}^{\infty} M^j (J \otimes H)(M')^j\} .$$

Let $z' = [z_1' \ldots z_n']$ where the z_i are p×1 vectors and $z \neq 0$. If $z_n \neq 0$, then $z'(J \otimes H)z = z_n'Hz_n > 0$ since H is positive definite. Thus $z'Vz \geq z'(J \otimes H)z > 0$. If $z_n = 0$, then there is an integer $j < n$ such that $z_j \neq 0$ but $z_i = 0$ for $i > j$. Now, because of the form of M,

$$z'M = [0', z_1', \ldots, z_j', 0', \ldots, 0']$$

and so

$$z'M^{n-j} = [0, \ldots, 0, z_1', \ldots, z_j'] .$$

However, $z'Vz \geq z'M^{n-j}(J \otimes H)(M')^{n-j}z = z_j'Hz_j > 0$. Thus $z'Vz > 0$ for all vectors z, and V is positive definite. Finally, the alternate form for H which is stated in the theorem is derived. From above

$$\text{vec } V = (I-M \otimes M)^{-1}\text{vec}(J \otimes H)$$

$$= A \text{ vec } H.$$

Thus C vec V = CA vec H. But vec H = vec G + C vec V = vec G + CA vec H, and vec H = $(I-CA)^{-1}$vec G, provided that (I-CA) is invertible. To see that (I-CA) is indeed invertible, suppose CA has a unit eigenvalue with left eigenvector z'. Then $z'CA = z'$. Defining the $n^2p^2 \times 1$ vector ω by

$$\omega' = z'C(I-M \otimes M)^{-1}$$

it follows that $\omega'\tilde{C} = z'C(I-M \otimes M)^{-1}\tilde{C} = z'CAC = z'C$. Thus $\omega'\tilde{C}(I-M \otimes M)^{-1} = z'C(I-M \otimes M)^{-1} = \omega'$, and

$$\omega'(I-M \otimes M-\tilde{C}) = 0.$$

Since $(M \otimes M+\tilde{C})$ has no unit eigenvalues, ω must therefore be the zero vector. But $z' = z'CA = \omega'\tilde{C}A = 0$, and $(I-CA)$ is invertible. #

It is worth noting that the matrices V and H may be calculated more easily than indicated in the above theorem. Again, letting $h = H_{np}$ and $k = K_{np}$, from (2.3.7) V satisfies

$$\text{vech } V = \text{vech}(J \otimes H) + (h(M \otimes M)k') \text{ vech } V$$

or

$$\text{vech } V = \{I-h(M \otimes M)k'\}^{-1}\text{vech}(J \otimes H) .$$

The vector $\text{vech}(J \otimes H)$ has at most its last $p(p+1)/2$ elements non-zero. Hence letting $\tilde{A}$ be the $np(np+1)/2 \times p(p+1)/2$ matrix formed from the last $p(p+1)/2$ columns of $\{I-h(M \otimes M)k'\}^{-1}$ we see that

$$\text{vech } V = \tilde{A} \text{ vech } H.$$

Also, $\text{vech } H = \text{vech } G + \tilde{h}C\tilde{k}' \text{ vech } V = \text{vech } G + \tilde{h}C\tilde{k}'\tilde{A} \text{ vech } H$, where $\tilde{h} = H_p$ and $\tilde{k} = K_p$. Thus vech H may be calculated from

$$\text{vech } H = (I-\tilde{h}C\tilde{k}'\tilde{A})^{-1} \text{ vech } G.$$

The following corollary obtains the conditions for the existence of a singly-infinite stationary solution $\{X(t);\ t = 1,2,\ldots\}$ to (1.1.1) deferred from section (2.2).

COROLLARY 2.3.1. $\{X(t);\ t = 1-n,\ldots,0,1,\ldots\}$ <u>generated by</u> (1.1.1) <u>is stationary and satisfies</u> (iv) <u>under the same conditions as the conditions of Theorem</u> 2.3, <u>provided that</u> $(M \otimes M+\tilde{C})$ <u>have no unit eigenvalues. If</u> $(M \otimes M+\tilde{C})$

does have an eigenvalue equal to unity, then a solution exists if and only if a positive definite matrix V exists which satisfies

$$\text{vec } V = (M \otimes M + \tilde{C}) \text{ vec } V + \text{vec}(J \otimes G) .$$

Proof. The proof follows directly from the corollary to theorem 2.1 and theorem 2.3. #

A very simple set of conditions results when $p = 1$, the same result being obtained by Andel (1976) for the existence of a singly-infinite solution to (1.1.1).

COROLLARY 2.3.2. *When $p = 1$ and $G > 0$ there exists a unique F_t-measurable stationary solution to (1.1.1) if and only if M has all its eigenvalues within the unit circle and $CA < 1$.*

Proof. Since G and K are scalars and $H = (I-CA)^{-1}G$, we have $H > 0$ only when $CA < 1$, and the result follows from theorem 2.3. #

As an example of the use of corollary 2.3.2, consider the case $n = p = 1$, and let $\beta = \beta_1$, $B_t = B_1(t)$ and $E(B_t^2) = \delta^2 = C$. Then the matrix M is equal to β, so that we must have $|\beta| < 1$ if there is to exist an F_t-measurable stationary solution to (1.1.1). Furthermore, $(I-M \otimes M)^{-1} = (1-\beta^2)^{-1}$ so that $CA = \delta^2/(1-\beta^2)$. Thus an F_t-measurable stationary solution exists to (1.1.1) if and only if $|\beta| < 1$ and $\delta^2 < 1 - \beta^2$, that is, if and only if $\beta^2 + \delta^2 < 1$, since this latter condition implies that $|\beta| < 1$.

2.4 The Case of a Unit Eigenvalue

Theorem 2.3 does not cover the case where $(M \otimes M+\tilde{C})$ possesses a unit eigenvalue. The main reason that this case has not been considered is that the matrix $(I-M \otimes M-\tilde{C})$ is not invertible if this is so. Another reason is provided in the proof of lemma 2.2, namely, a solution, if it exists, may not be unique. As seen in corollary 2.2.2, however, an F_t-measurable stationary solution exists, in the case where $p = 1$ and $G \neq 0$, if and only if <u>all</u> the eigenvalues of $(M \otimes M+\tilde{C})$ have moduli less than unity, so that in the univariate case, the difficulty does not arise.

When $p > 1$ it is possible that processes $\{X(t), t = 1,2,...\}$ exist which satisfy (1.1.1), are stationary, satisfy conditions (i)-(iv) and for which $(M \otimes M+C)$ has a unit eigenvalue. To see this, let $\{X(t), t = 1,2,...\}$ be such that

$$X(t) = (\beta + B(t))X(t-1) + \varepsilon(t)$$

where $X(t)$ and $\varepsilon(t)$ are 2×1 random vectors (i.e. $p = 2$ in (1.1.1)), and

$$\beta = \begin{bmatrix} b & 0 \\ 0 & 0 \end{bmatrix}, \quad B(t) = \begin{bmatrix} 0 & 0 \\ 0 & \beta(t) \end{bmatrix}, \quad \varepsilon(t) = \begin{pmatrix} \delta(t) \\ 0 \end{pmatrix},$$

with $E(\delta(t)) = E(\beta(t)) = 0$, $E(\beta^2(t)) = 0$, $E\delta^2(t) = g$, $|b| < 1$ and $\delta(t)$, $\beta(t)$ are independent. Furthermore, let $E(X(0)) = 0$ and

$$E[X(0)X'(0)] = [g/(1-b^2),\ 0,\ 0,\ c]' \quad \text{where } c > 0$$

Then $\text{vec}(J \otimes G) = \text{vec } G = g[1,0,0,0]'$, and the matrix $M = \beta$ has eigenvalues 0 and 1, while

$$(M \otimes M+\tilde{C}) = \begin{bmatrix} b^2 & 0 & 0 & 0 \\ 0 & 0 & 0 & 0 \\ 0 & 0 & 0 & 0 \\ 0 & 0 & 0 & 1 \end{bmatrix}$$

has eigenvalues 0, 0, b^2 and 1. Now,

$$\begin{aligned} \text{vec } E(X(1)X'(1)) &= (M \otimes M+\tilde{C}) \text{ vec } E(X(0)X'(0)) + \text{vec}(J \otimes G) \\ &= [gb^2/(1-b^2)\ 0\ 0\ c]' + [g\ 0\ 0\ 0]' \\ &= [g/(1-b^2)\ 0\ 0\ c]' \\ &= \text{vec } E(X(0)X'(0)). \end{aligned}$$

Also, $E(X(1) = \beta E(X(0)) = 0 = EX(0)$. Hence $\{X(t); t = 1,2,...\}$ is stationary by theorem 2.1. Noting that c is arbitrary, the number of such stationary solutions is seen to be uncountable. #

2.5 Stability of RCA Models

When generating a time series by an equation such as (1.1.1), it is usual to initialise the variables $\{X(1-n), X(2-n), \ldots, X(-1), X(0)\}$ and commence the generation at $t = 1$. An obvious question to ask is whether or not these initial values affect the long-term behaviour of the process $\{X(t);\ t = 1,2,\ldots\}$, and whether the process attains some equilibrium as t becomes large, a question which is of importance to econometricians when considering economic systems. The concept of stability, introduced in definition 2.1.1, provides a concrete way in which to frame this question.

The following theorem will prove useful in obtaining the eventual necessary and sufficient conditions for stability.

THEOREM 2.4. $\{X(t);\ t = 1,2,\ldots\}$ generated by (1.1.1) is stable if and only if $(M \otimes M+\tilde{C})^t$ vec S converges to zero for all symmetric $np \times np$ matrices S as $t \to \infty$.

Proof. Letting $y(0) = [x'(1-n), \ldots, x'(0)]'$ it is seen from (2.2.1) that

$$\begin{aligned} E\{Y(t) \mid Y(0) = y(0)\} &= ME\{Y(t-1) \mid Y(0) = y(0)\} \\ &= M^t y(0) \end{aligned}$$

and so $E\{Y(t) \mid Y(0) = y(0)\}$ converges independently of $y(0)$ if and only if all the eigenvalues of M have moduli less than unity. For if M had an eigenvalue whose modulus were greater than or equal to unity, $M^t z$ would not converge at all if z were the corresponding right eigenvector, or its real or imaginary part. The only exception is where M has all unit eigenvalues. In this case however $M^t z$ will not converge for all z unless $M = I$, which is impossible.

Let $V_{t,t-s}(y(0)) = E(Y(t)Y'(t-s) \mid Y(0) = y(0))$, $t > s \geq 0$. From (2.2.1), we have for $s > 0$,

$$
\begin{aligned}
V_{t,t-s}(y(0)) &= E([\{M+D(t)\}Y(t-1)+\eta(t)]Y'(t-s) \mid Y(0) = y(0)) \\
&= ME(Y(t-1)Y'(t-s) \mid Y(0) = y(0)) \\
&= MV_{t-1,t-s}(y(0)) \\
&= M^s V_{t-s,t-s}(y(0)).
\end{aligned}
$$

Again, using (2.2.1) and essentially the same derivation as used in the proof of theorem 2.1,

$$
\begin{aligned}
\text{(2.5.1)} \qquad \operatorname{vec} V_{t,t}(y(0)) &= (M \otimes M+\tilde{C})^j \operatorname{vec} V_{t-1,t-1}(y(0)) + \operatorname{vec}(J \otimes G) \\
&= \sum_{j=0}^{t-1} (M \otimes M+\tilde{C})^j \operatorname{vec}(J \otimes G) + (M \otimes M+\tilde{C})^t \operatorname{vec}(y(0)y'(0)) .
\end{aligned}
$$

Now each of the terms on the right hand side of (2.5.1) is the vec of a non-negative definite matrix. Hence $V_{t,t}(y(0))$ converges if and only if both $(M \otimes M+\tilde{C})^t \operatorname{vec}(y(0)y'(0))$ and $\sum_{j=0}^{t-1} (M \otimes M+\tilde{C})^j \operatorname{vec}(J \otimes G)$ converge.

We may now show the sufficiency of the condition.

If $(M \otimes M+\tilde{C})^t \operatorname{vec} S$ converges to zero for all symmetric matrices S, then $(M \otimes M+\tilde{C})^t \operatorname{vec}(y(0)y'(0))$ and $(M \otimes M+\tilde{C})^t \operatorname{vec}(J \otimes G)$ converge to zero. Furthermore, they converge to zero at a geometric rate determined by some eigenvalue of $(M \otimes M+\tilde{C})$ less than unity in modulus, since $(M \otimes M+\tilde{C})^t = P\Lambda^t P^{-1}$, where $P\Lambda P^{-1}$ is a Jordan canonical representation of $(M \otimes M+\tilde{C})$. However, Λ^t has the tth powers of the eigenvalues of $(M \otimes M+\tilde{C})$ down its diagonal. Thus $\sum_{j=1}^{t-1} (M \otimes M+\tilde{C})^j \operatorname{vec}(J \otimes G)$ converges, and M has all its eigenvalues inside the unit circle, which follows from an application of lemma 2.1 in which the matrix G is replaced with any positive definite matrix Ω. Hence, since $V_{t,t-s}(y(0)) = M^s V_{t-s,t-s}(y(0))$ also converges, the condition is seen to be sufficient for stability.

To see that the condition is also necessary, note that any real symmetric $np \times np$ matrix S may be rewritten as $\sum_{j=1}^{np} \lambda_j e_j e_j'$, where $\{(\lambda_j, e_j)\}$ is the set of eigenvalues and corresponding right eigenvectors of S. Since $(M \otimes M+\tilde{C})^t \operatorname{vec}(yy')$ converges to zero for all $y \in \mathbb{R}^{np}$, so must

$$
\sum_{j=1}^{np} \lambda_j (M \otimes M+\tilde{C})^t \operatorname{vec}(e_j e_j') = (M \otimes M+\tilde{C})^t \operatorname{vec} S. \qquad \#
$$

The following necessary and sufficient condition is now easily obtained.

THEOREM 2.5. *$\{X(t);\ t = 1,2,\ldots\}$ generated by (1.1.1) is stable if and only if all the eigenvalues of $(h(M \otimes M+\tilde{C})k')$ have moduli less than unity, where $h = H_{np}$ and $k = K_{np}$.*

Proof. Let $W_t(S)$ be the $np \times np$ matrix defined by $\mathrm{vec}\, W_t(S) = (M \otimes M+\tilde{C})^t \mathrm{vec}\, S$. Then, since $\mathrm{vec}\, S = k'\, \mathrm{vech}\, S$,

$$\begin{aligned}\mathrm{vech}\, W_t(S) &= h\, \mathrm{vec}\, W_t(S)\\ &= h\,(M \otimes M+\tilde{C})k'\, \mathrm{vech}\, W_{t-1}(S)\\ &= [h(M \otimes M+\tilde{C})k']^t\, \mathrm{vech}\, S,\end{aligned}$$

by induction. Now, the vectors vech S formed from all symmetric $np \times np$ matrices S span all of $\mathbb{R}^{np(np+1)/2}$, so that $\mathrm{vech}\, W_t(S)$ converges to zero for all symmetric matrices S if and only if all the eigenvalues of $[h(M \otimes M+\tilde{C})k']$ have moduli less than unity. #

The necessary and sufficient condition derived by Conlisk (1974) is that all the eigenvalues of the matrix $(M \otimes M+\tilde{C})$ have moduli less than unity. While the condition is correct $(M \otimes M+\tilde{C})$ has n^2p^2 eigenvalues, while $[h(M \otimes M+\tilde{C})k']$ has only $np(np+1)/2$ eigenvalues. Since the computation of eigenvalues is costly in terms of computing time, the condition derived here is obviously more appealing from a practical point of view.

There is evidently a link between stability and stationarity. The relationship between these two concepts for random coefficient autoregressions is made clear in the following theorem.

THEOREM 2.6. *If $\{X(t);\ t = 1,2,\ldots\}$ generated by (1.1.1) is stable, there exists a stationary F_t-measurable solution $\{X^*(t)\}$ to (1.1.1) for which $E[(X(t)-X^*(t)) \mid Y(0) = y(0)]$ and $E[(X(t)-X^*(t))(X(t-s)-X^*(t-s))' \mid Y(0) = y(0)]$, for fixed s, converge to zero as t increases.*

Proof. By corollary 2.2.1, there exists a unique F_t-measurable stationary solution $\{X^*(t)\}$ to (1.1.1). Letting $Y^*(t) = [X'(t+1-n),\ldots,X^{*\prime}(t)]'$, $Y^*(t)$ is given by (2.3.2). Now, putting $Y(0) = y(0)$ in (2.3.1), we have

$$Y(t) = \sum_{j=0}^{t-1} S_{t,j-1}\eta(t-j) + S_{t,t-1}y(0).$$

Thus

$$Y(t)-Y^*(t) = S_{t,t-1}y(0) - \sum_{j=t}^{\infty} S_{t,j-1}\eta(t-j) \tag{2.5.2}$$

and

$$E[(Y(t)-Y^*(t)) \mid Y(0) = y(0)] = M^t y(0)$$

which converges to zero since the eigenvalues of M have moduli less than unity. Also

$$E[(Y(t)-Y^*(t))(Y(t-s)-Y^*(t-s))' \mid Y(0) = y(0)]$$
$$= E[(M+D(t))(Y(t-1)-Y^*(t-1))(Y(t-s)-Y^*(t-s))' \mid Y(0) = y(0)]$$
$$= ME[(Y(t-1)-Y^*(t-1))(T(t-s)-Y^*(t-s))' \mid Y(0) = y(0)]$$
$$= M^s E[(Y(t-s)-Y^*(t-s))(Y(t-s)-Y^*(t-s))' \mid Y(0) = y(0)].$$

Since $\{\eta(t)\}$ is independent of $\{D(t)\}$, the two terms in the right hand side of (2.5.2) are uncorrelated. Thus

$$\text{vec } E[(Y(t)-Y^*(t))(Y(t)-Y^*(t))' \mid Y(0) = y(0)]$$

$$= E[(\prod_{j=0}^{t-1} \{M+D(t-j)\} \otimes \prod_{k=0}^{t-1} \{M+D(t-k)\})\text{vec}(y(0)y'(0))]$$

$$+ E[\sum_{j=t}^{\infty} (\prod_{k=0}^{j-1} \{M+D(t-k)\} \otimes \prod_{\ell=0}^{j-1} \{(M+D(t-\ell)\})\text{vec}(\eta(t-j)\eta'(t-j))]$$

$$= E[\prod_{j=0}^{t-1} (\{M+D(t-j)\} \otimes \{M+D(t-j)\})]\text{vec}(y(0)y'(0))$$

$$+ E[\sum_{j=t}^{\infty} \{ \prod_{k=0}^{j-1} (\{M+D(t-k)\} \otimes \{M+D(t-k)\})\}]E\ \text{vec}(\eta(t-j)\eta'(t-j))$$

$$= (M \otimes M+\tilde{C})^t \text{vec}(y(0)y'(0)) + \sum_{j=t}^{\infty} (M \otimes M+\tilde{C})^j \text{vec}(J \otimes G)$$

since $\{\eta(t)\}$ and $\{D(t)\}$ are both uncorrelated processes. Hence

$E[(Y(t)-Y^*(t))(Y(t-s)-Y^*(t-s))' \mid Y(0) = y(0)]$ converges to zero for all s since $(M \otimes M+\tilde{C})^t \mathrm{vec}(y(0)y'(0))$ converges to zero and $\sum_{j=0}^{t-1} (M \otimes M+\tilde{C})^j \mathrm{vec}(J \otimes G)$ converges, its tail sum $\sum_{j=t}^{\infty} (M \otimes M+\tilde{C})^j \mathrm{vec}(J \otimes G)$ thus converging to zero. #

2.6 Strict Stationarity

The previous sections have assumed nothing about $\{\varepsilon(t)\}$ and $\{B(t)\}$ except that they are independent second order stationary processes which are mutually independent.

If $\{\varepsilon(t)\}$ and $\{B(t)\}$ are also sequences of identically distributed random variables, in which case they are also strictly stationary and ergodic, it is possible to infer stronger properties for the F_t-measurable solution $\{X(t)\}$ to (1.1.1), properties which are required in later chapters.

THEOREM 2.7. *Suppose $\{\varepsilon(t)\}$ and $\{B(t)\}$ satisfy assumptions* (i) *and* (iii) *and are also identically distributed sequences. Then, if a unique F_t-measurable second order stationary solution $\{X(t)\}$ exists to* (1.1.1), $\{X(t)\}$ *is also strictly stationary and ergodic.*

Proof. The unique second order stationary F_t-measurable solution $Y(t)$ to (2.2.1), given by (2.3.2) is the limit in mean square, and hence in probability, of a sequence of F_t-measurable random variables. Since the solution has the same functional form for each t, $\{Y(t)\}$ must therefore be strictly stationary, as must $\{X(t)\}$. Now, $\{(\varepsilon(t),B(t))\}$ is an ergodic sequence since it is a sequence of independent, identically distributed random vectors. Also, the σ-field G_t generated by $\{X(t),X(t-1),\ldots\}$ is a subset of F_t if $\{X(t)\}$ is an F_t-measurable sequence of random variables. Letting G and F be the smallest σ-fields containing $\lim_{t\to\infty} G_t$ and $\lim_{t\to\infty} F_t$ respectively, it follows that $G \subset F$ and $\{X(t)\}$ is ergodic.

APPENDIX 2.1

PROOF OF LEMMA 2.1. Define the matrix W by

$$\text{vec } W = \sum_{j=0}^{\infty} (M \otimes M + C)^j \text{vec}(J \otimes G) .$$

Then

$$(M \otimes M + \tilde{C}) \text{ vec } W = \sum_{j=1}^{\infty} (M \otimes M + \tilde{C})^j \text{vec}(J \otimes G)$$

$$= \text{vec } W - \text{vec}(J \otimes G)$$

and

$$\text{vec } W = (M \otimes M)\text{vec } W + (\tilde{C} \text{ vec } W + \text{vec}(J \otimes G))$$

$$= (M \otimes M) \text{ vec } W + \text{vec}(J \otimes H)$$

since $\tilde{C}$ vec W = vec$(J \otimes (C \text{ vec } W))$, because of the positions taken by the only possible non-zero elements of the vector $\tilde{C}$ vec W. Hence $W = MWM' + J \otimes H$.

Let λ be an eigenvalue of M, with corresponding left eigenvector $z' \neq 0$ where $z' = [z_1' \ldots z_n']$ and the z_i are $p \times 1$ vectors. Then

$$z'W\bar{z} = z'MWM'\bar{z} + z'(J \otimes H)\bar{z}$$

$$= |\lambda|^2 z'W\bar{z} + z_n'H\bar{z}_n .$$

That is, $(1-|\lambda|^2)z'W\bar{z} = z_n'H\bar{z}_n$.

Now, as seen before, the term $\sum_{j=0}^{r} (M \otimes M+\tilde{C})^j \text{vec}(J \otimes G)$ is the vec of a non-negative definite matrix, so that the limit W is also non-negative definite, and $z'W\bar{z} \geq 0$. If $z_n'H\bar{z}_n > 0$, that is, if $z_n \neq 0$, we have $|\lambda| < 1$. Suppose now that $z_n = 0$. Then, since z' is a left eigenvector of M,

$$[z_1' \ldots z_n'] \left[\begin{array}{c:c} 0 & I \\ \hdashline \beta_n \cdots\cdots & \beta_1 \end{array}\right] = \lambda[z_1' \ldots z_n']$$

which reduces to the following set of equations

$$z_n'\beta_n = \lambda z_1' ,$$

$$z_i' + z_n'\beta_{n-i} = \lambda z_{i+1}' , \quad i = 1,\ldots,n-1 .$$

If $\lambda \neq 0$, the first equation gives $z_1 = 0$, since $z_n = 0$, and the remaining equations have as their only solution $z_2 = \ldots = z_{n-1} = 0$. However, $z \neq 0$ so that in any case, we must have $|\lambda| < 1$. #

PROOF OF LEMMA 2.2. Suppose there are two solutions to (2.2.1), $W(t)$ and $Z(t)$, and let $U(t) = W(t) - Z(t)$. Then $U(t)$ satisfies

$$U(t) = (M+D(t))U(t-1)$$

and, since $U(t)$ is also F_t-measurable,

$$\text{vec } E(U(t)U'(t)) = (M \otimes M+\tilde{C})\text{vec } E(U(t-1)U'(t-1)) .$$

However, since $\{U(t)\}$ is also stationary we must have

$$\text{vec } E(U(t)U'(t)) = (M \otimes M+\tilde{C})\text{vec } E(U(t)U'(t))$$

and $E(U(t)U'(t)) = 0$ since $(M \otimes M+\tilde{C})$ has no unit eigenvalues. Thus $U(t) = 0$ and $W(t) = Z(t)$ almost everywhere. #

CHAPTER 3

LEAST SQUARES ESTIMATION OF SCALAR MODELS

3.1 Introduction

In chapter 2, conditions were found for the existence of stationary solutions to equations of the form (1.1.1). In practice, however, given that a stationary time series $\{X(t)\}$ satisfies such an equation, it is necessary to estimate the unknown parameters in order to provide predictors of $X(t)$ given past values of the process. Estimation procedures for fixed coefficient autoregressions are well established, and the asymptotic properties of these estimates are well known (see, for example, chapter 6 of Hannan (1970)). Random coefficient autoregressions are, however, non-linear in nature, and any foreseeable maximum likelihood type estimation method would be an iterative procedure. Such a procedure is discussed in Chapter 4, where the asymptotic properties of the estimates obtained are determined. Iteration must, nevertheless, commence at some point, and since the likelihood will be non-linear and its domain will be of relatively high dimensions, it is likely that there will be local extrema. Hence it is desirable that iterations commence close to the global maximum of the likelihood function for otherwise convergence might be toward a local extremum. In this chapter, a least squares estimation procedure is proposed for univariate random coefficient autoregressions which, under certain conditions, is shown to give strongly consistent estimates of the true parameters. The estimates are also shown to obey a central limit theorem. It is these least squares estimates which will be used to commence the iterative procedure which optimizes the likelihood criterion to be considered in the next chapter.

As well as the conditions (i)-(iv) assumed in chapter 2, we shall make the further assumptions

(v) $\{\varepsilon(t)\}$ and $\{B(t)\}$ are each identically distributed sequences.

(vi) The parameters β_i, $i = 1,\ldots,n$ and C are such that a unique second order stationary F_t-measurable solution $\{X(t)\}$ to (2.1.1) exists.

We emphasize that only scalar models, that is, models with $p = 1$, are considered in this and the following chapter. The generalization to multivariate models is discussed in chapter 7. Also, by assuming (v) and (vi), theorem 2.7 shows that the solution $\{X(t)\}$ is strictly stationary and ergodic, since it is unique by corollary 2.2.2 and lemma 2.2.

It will prove necessary to make a further assumption concerning $\{X(t)\}$. Letting $z(t) = K_n \text{vec}\{Y(t-1)Y'(t-1)\}$, where K_n is defined in appendix A.1, we shall need to know that there is no $n(n+1)/2$-component vector α such that $\alpha'(z(t)-E[z(t)]) = 0$ almost everywhere. This will be proved in lemma 3.1 under conditions (i)-(v) along with the condition

(vii) $\varepsilon(t)$ cannot take on only two values almost surely.

Letting $\sigma^2 = G \neq 0$ and $\Sigma = E[B'(t)B(t)]$, it is easily seen that $\text{vec}\,\Sigma = E(B'(t) \otimes B'(t)) = \{E(B(t) \otimes B(t))\}' = C'$, since $p=1$. Now, from corollary 2.3.2, the necessary and sufficient conditions that condition (vi) hold are that M have all its eigenvalues inside the unit circle, or equivalently that

$1 - \sum_{i=1}^{n} \beta_i z^i$ have all its zeros outside the unit circle, which is shown

in Andel (1971), and that CA be less than unity, where A is the last column of $(I-M \otimes M)^{-1}$. Letting W be the $n\times n$ matrix for which $A = \text{vec}\,W$, this latter condition may be replaced by the condition that $\text{tr}(\Sigma W) < 1$, since $CA = (\text{vec}\,\Sigma)'\text{vec}\,W = \text{tr}(\Sigma W)$.

3.2 The Estimation Procedure

The estimation procedure is a generalization of a two-step procedure proposed by Rosenberg (1973) to estimate the parameters of a random coefficient regression model. Since the matrix Σ is symmetric, we need only estimate $\gamma = \text{vech}\,\Sigma$. The first step is to estimate the parameters β_i, $i = 1,\dots,n$. From (1.1.1),

$$X(t) = \sum_{i=1}^{n} \beta_i X(t-i) + \sum_{i=1}^{n} B_i(t)X(t-i) + \varepsilon(t)$$

or

$$X(t) = \beta' Y(t-1) + u(t) \ , \qquad (3.2.1)$$

where $\beta = [\beta_n \ \dots \ \beta_1]'$ and $u(t) = B(t)Y(t-1) + \varepsilon(t)$. Letting F_t be the σ-field generated by $\{(\varepsilon(t),B(t)), (\varepsilon(t-1),B(t-1)),\dots\}$, we have

$$\begin{aligned} E(u(t)|F_{t-1}) &= E\{B(t)\}Y(t-1) + E\{\varepsilon(t)\} \\ &= 0 \end{aligned}$$

since $B(t)$ and $\varepsilon(t)$ are independent of $\{(\varepsilon(t-1),B(t-1)),(\varepsilon(t-2),B(t-2)),\dots\}$ and $Y(t-1)$ is a measurable function of this set alone. Also,

$$\begin{aligned} &E(u^2(t)|F_{t-1}) \\ &\quad = E\{\varepsilon^2(t)\} + 2E\{\varepsilon(t)B(t)Y(t-1)|F_{t-1}\} + E\{[B(t)Y(t-1)]^2|F_{t-1}\} \\ &\quad = \sigma^2 + 2E\{\varepsilon(t)\}E\{B(t)Y(t-1)|F_{t-1}\} + E\{Y'(t-1)B'(t)B(t)Y(t-1)|F_{t-1}\} \\ &\quad = \sigma^2 + Y'(t-1)E\{B'(t)B(t)\}Y(t-1) = \sigma^2 + Y'(t-1)\Sigma Y(t-1) \\ &\quad = \sigma^2 + \{Y'(t-1) \otimes Y'(t-1)\}\text{vec}\,\Sigma = \sigma^2 + \{\text{vec}[Y(t-1)Y'(t-1)]\}'K_n'\,\text{vech}\,\Sigma \ . \end{aligned}$$

That is,

$$E(u^2(t)|F_{t-1}) = \sigma^2 + z'(t)\gamma = \sigma^2 + \gamma' z(t) \qquad (3.2.2)$$

where $\gamma = \text{vech}\,\Sigma$ and $z(t) = K_n\{\text{vec}[Y(t-1)Y'(t-1)]\}$. Given the sample $\{X(1-n),\dots,X(0),X(1),\dots,X(N)\}$, we obtain the least squares estimate $\hat{\beta}$ of β from (3.2.1) by minimizing $\sum_{t=1}^{N} u^2(t)$ with respect to β. Thus $\hat{\beta}$ is given by

$$(3.2.3) \qquad \hat{\beta} = \left\{ \sum_{t=1}^{N} Y(t-1)Y'(t-1) \right\}^{-1} \sum_{t=1}^{N} Y(t-1)X(t) .$$

The second step in the estimation procedure begins by using (3.2.1) to form the residuals $\hat{u}(t) = X(t) - \hat{\beta}'Y(t-1)$, $t = 1,\ldots,N$. In view of (3.2.2), let $\eta(t) = u^2(t) - \sigma^2 - z'(t)\gamma$. Then the estimates $\hat{\gamma}$ and $\hat{\sigma}^2$ of γ and σ^2 respectively are obtained by minimizing $\sum_{t=1}^{N} \eta^2(t)$ with respect to γ and σ^2, that is, by regressing $\hat{u}^2(t)$ on 1 and $z(t)$. Thus

$$(3.2.4) \qquad \hat{\gamma} = \left\{ \sum_{t=1}^{N} (z(t)-\bar{z})(z(t)-\bar{z})' \right\}^{-1} \sum_{t=1}^{N} \hat{u}^2(t)(z(t)-\bar{z})$$

and

$$(3.2.5) \qquad \hat{\sigma}^2 = N^{-1} \sum_{t=1}^{N} \hat{u}^2(t) - \hat{\gamma}'\bar{z}$$

where $\bar{z} = N^{-1} \sum_{t=1}^{N} z(t)$.

It should be noted that $N^{-1} \sum_{t=1}^{N} Y(t-1)Y'(t-1)$ is positive definite almost everywhere for large enough N, for otherwise there would exist an n-component non-zero vector α with $\alpha'Y(t-1) = 0$, $t = 1,2,\ldots$, which is precluded since $V = E(Y(t-1)Y'(t-1))$ was shown to be positive definite under conditions (i)-(vi) in chapter 2. Also, $N^{-1} \sum_{t=1}^{N} (z(t)-\bar{z})(z(t)-\bar{z})'$ is positive definite almost everywhere for large enough N, since it will be shown in lemma 3.1 that there is no $n(n+1)/2$-component non-zero vector α such that $\alpha'(z(t)-E(z(t))) = 0$, almost surely, and since $\bar{z}$ converges to $E(z(t))$ by the ergodic theorem.

Equations (3.2.3)-(3.2.5) define the least squares estimates $\hat{\beta}, \hat{\gamma}$ and $\hat{\sigma}^2$ of the parameters β, γ and σ^2.

3.3 Strong Consistency and the Central Limit Theorem

The strong consistency of the estimates defined by (3.2.3)-(3.2.5) will be shown using the ergodic theorem, while Billingsley's martingale central limit theorem (theorem A.1.4) will be used to provide the central limit theorem. It is convenient firstly to obtain the results for $\hat{\beta}$, this being required since the residuals $\hat{u}(t)$ used to derive the estimates $\hat{\sigma}^2$ and $\hat{\gamma}$ are not the true residuals.

THEOREM 3.1: For a strictly stationary F_t-measurable process $\{X(t)\}$ satisfying (1.1.1) under assumptions (i)-(vi) with $p = 1$, and $\hat{\beta}$ given by (3.2.3), $\hat{\beta}$ converges almost surely to β. Furthermore, if $E(X^4(t)) < \infty$, then $N^{\frac{1}{2}}(\hat{\beta}-\beta)$ has a distribution which converges to the normal distribution with mean zero and covariance matrix $\sigma^2 V^{-1} + V^{-1}E[Y(t-1)Y'(t-1)\gamma' z(t)]V^{-1}$, where $V = E[Y(t-1)Y'(t-1)]$.

From (3.2.3),

$$\begin{aligned}
\hat{\beta}-\beta &= \{N^{-1}\sum_{t=1}^{N} Y(t-1)Y'(t-1)\}^{-1}\{N^{-1}\sum_{t=1}^{N} Y(t-1)X(t)\} - \beta \\
&= \{N^{-1}\sum_{t=1}^{N} Y(t-1)Y'(t-1)\}^{-1} N^{-1}\sum_{t=1}^{N} \{Y(t-1)X(t)-Y(t-1)Y'(t-1)\beta\} \\
&= \{N^{-1}\sum_{t=1}^{N} Y(t-1)Y'(t-1)\}^{-1} N^{-1}\sum_{t=1}^{N} Y(t-1)u(t) .
\end{aligned}$$

Since $\{X(t)\}$ is strictly stationary and ergodic, so are $\{Y(t)Y'(t)\}$ and $\{Y(t-1)u(t)\}$. Furthermore, $V = E\{Y(t)Y'(t)\}$ is finite by (vi) and $E\{Y(t-1)u(t)\} = E\{E[Y(t-1)u(t)|F_{t-1}]\} = E\{Y(t-1)E(u(t)|F_{t-1})\} = 0$, since $E(u(t)|F_{t-1}) = 0$, and $Y(t-1)$ is a measurable function of $\{(\varepsilon(t-1),B(t-1)), (\varepsilon(t-2),B(t-2)),\ldots\}$ alone. Thus $N^{-1}\sum_{t=1}^{N} Y(t-1)Y'(t-1)$ converges almost surely to V, and $N^{-1}\sum_{t=1}^{N} Y(t-1)u(t)$ converges almost surely to zero, showing that $(\hat{\beta}-\beta)$ converges almost surely to 0.

Now, if α is any n-component vector,

$$E\{(\alpha'Y(t-1)u(t))^2\} = E\{E(\alpha'Y(t-1)u(t))^2 | F_{t-1}\}$$
$$= E\{(\alpha'Y(t-1))^2 E(u^2(t) | F_{t-1})\}$$
$$= E\{(\alpha'Y(t-1))^2 (\sigma^2 + \gamma'z(t))\}$$

by (3.2.2), the expectation existing if $E(X^4(t)) < \infty$ since the components of $(\alpha'Y(t-1))^2(\gamma'z(t))$ are quartic in $\{X(t)\}$. Since $E\{\alpha'Y(t-1)u(t) | F_{t-1}\} = 0$, an application of theorem A.1.4 shows that $N^{-\frac{1}{2}} \sum_{t=1}^{N} (\alpha'Y(t-1))u(t)$ has a distribution which converges to the normal distribution with mean zero and variance

$$E\{(\alpha'Y(t-1))^2(\sigma^2+\gamma'z(t))\} = \alpha'E\{Y(t-1)Y'(t-1)(\sigma^2+\gamma'z(t))\}\alpha$$

for all $\alpha \in \mathbb{R}^n$ provided that $E(X^4(t)) < \infty$. Thus $N^{-\frac{1}{2}} \sum_{t=1}^{N} Y(t-1)u(t)$ converges in distribution to the multivariate normal distribution with mean zero and covariance matrix $E\{Y(t-1)Y'(t-1)(\sigma^2+\gamma'z(t))\}$. Hence $N^{\frac{1}{2}}(\hat{\beta}-\beta)$ converges in distribution to the normal distribution with mean zero and covariance matrix

$$V^{-1}E\{Y(t-1)Y'(t-1)[\sigma^2+\gamma'z(t)]\}V^{-1} = \sigma^2V^{-1}VV^{-1}+V^{-1}E\{Y(t-1)Y'(t-1)\gamma'z(t)\}V^{-1}$$
$$= \sigma^2V^{-1}+V^{-1}E\{Y(t-1)Y'(t-1)\gamma'z(t)\}V^{-1} ,$$

using the fact that if a vector $\omega(N)$ converges in distribution to the normal distribution with mean zero and covariance matrix W, and a matrix $A(N)$ converges in probability to a matrix A, then $A(N)\omega(N)$ converges in distribution to the normal distribution with mean zero and covariance matrix AWA'. #

We shall need to have the matrix $E\{(z(t)-E[z(t)])(z(t)-E[z(t)])'\}$ positive definite if $E(X^4(t)) < \infty$, in order to derive the asymptotic behaviour of $\hat{\gamma}$ and $\hat{\sigma}^2$. This fact is guaranteed by the following lemma, the proof of which is given in appendix 3.1.

LEMMA 3.1: Under assumptions (i)-(vii), there is no non-zero $n(n+1)/2$-component vector α such that $\alpha'(z(t)-E[z(t)]) = 0$ almost everywhere for all t.

Before the proof of the main theorem is given, we shall need to establish a further lemma which examines the effect on the estimates $\hat{\sigma}^2$ and $\hat{\gamma}$ by having $u(t)$ in (3.2.4) and (3.2.5) instead of $\hat{u}(t)$. In order to do this, let $\tilde{\gamma}$ and $\tilde{\sigma}^2$ be given by

$$\text{(3.3.1)} \qquad \tilde{\gamma} = \left\{ \sum_{t=1}^{N} (z(t)-\bar{z})(z(t)-\bar{z})' \right\}^{-1} \sum_{t=1}^{N} u^2(t)(z(t)-\bar{z}) ,$$

$$\text{(3.3.2)} \qquad \tilde{\sigma}^2 = N^{-1} \sum_{t=1}^{N} u^2(t) - \tilde{\gamma}'\bar{z} .$$

LEMMA 3.2: $(\tilde{\gamma}-\hat{\gamma})$ and $(\tilde{\sigma}^2-\hat{\sigma}^2)$ converge almost surely to zero if $E(X^4(t)) < \infty$, while $N^{\frac{1}{2}}(\tilde{\gamma}-\hat{\gamma})$ and $N^{\frac{1}{2}}(\tilde{\sigma}^2-\hat{\sigma}^2)$ converge in probability to zero.

Proof: See appendix 3.1.

The derivation of the asymptotic properties of the estimate $\hat{K} = [\hat{\beta}', \hat{\gamma}', \hat{\sigma}^2]'$ of $K = [\beta', \gamma', \sigma^2]'$ is now straightforward.

THEOREM 3.2: For a strictly stationary F_t-measurable process $\{X(t)\}$ satisfying (1.1.1) under assumptions (i)-(vii) with $p = 1$ and $\hat{\beta}$, $\hat{\gamma}$ and $\hat{\sigma}^2$ given by (3.2.3)-(3.2.5), $\hat{K} = [\hat{\beta}', \hat{\gamma}', \hat{\sigma}^2]'$ converges almost surely to $K = [\beta', \gamma', \sigma^2]'$ if $E(X^4(t)) < \infty$, while if $E(X^8(t)) < \infty$ the distribution of $N^{\frac{1}{2}}(\hat{K}-K)$ converges to the normal distribution with mean zero and covariance matrix Ω defined by (3.3.9) and derived in appendix 3.2.

Proof: As a consequence of lemma 3.2, if $\tilde{K} = [\hat{\beta}', \tilde{\gamma}', \tilde{\sigma}^2]$, then $(\hat{K}-\tilde{K})$ converges almost surely to zero, while $N^{\frac{1}{2}}(\hat{K}-\tilde{K})$ converges in probability to zero if $E(X^4(t)) < \infty$. It will be shown that $(\tilde{K}-K)$ converges almost surely to zero, while $N^{1/2}(\tilde{K}-K)$ converges in distribution so that $(\hat{K}-K)$ converges almost surely to zero, while $N^{1/2}(\hat{K}-K)$ converges in distribution in the same way as $N^{1/2}(\tilde{K}-K)$. Thus we need only prove the results for $\tilde{K}$. From (3.3.1) we have

$$\begin{aligned}\tilde{\gamma}-\gamma &= \{N^{-1}\sum_{t=1}^{N}(z(t)-\bar{z})(z(t)-\bar{z})'\}^{-1} N^{-1}\sum_{t=1}^{N}(z(t)-\bar{z})u^2(t) - \gamma \\ &= \{N^{-1}\sum_{t=1}^{N}(z(t)-\bar{z})(z(t)-\bar{z})'\}^{-1} N^{-1}\sum_{t=1}^{N}(z(t)-\bar{z})\{u^2(t)-(z(t)-\bar{z})'\gamma\} \\ &= \{N^{-1}\sum_{t=1}^{N}(z(t)-\bar{z})(z(t)-\bar{z})'\}^{-1} N^{-1}\sum_{t=1}^{N}(z(t)-\bar{z})\{u^2(t)-z'(t)\gamma\}\end{aligned}$$

since $\sum_{t=1}^{N}(z(t)-\bar{z})\bar{z}'\gamma = \left\{\sum_{t=1}^{N}(z(t)-\bar{z}\right\}\bar{z}'\gamma = 0$.

Letting $\xi(t) = u^2(t) - \sigma^2 - \gamma' z(t)$, we have, from the above,

$$\tilde{\gamma}-\gamma = \{N^{-1}\sum_{t=1}^{N}(z(t)-\bar{z})(z(t)-\bar{z})'\}^{-1} N^{-1}\sum_{t=1}^{N}(z(t)-\bar{z})\xi(t) \tag{3.3.7}$$

since $\sum_{t=1}^{N}(z(t)-\bar{z})\sigma^2 = 0$. Also, from (3.3.1),

$$\begin{aligned}\tilde{\sigma}^2-\sigma^2 &= N^{-1}\sum_{t=1}^{N} u^2(t)-\tilde{\gamma}'\bar{z}-\sigma^2 \\ &= N^{-1}\sum_{t=1}^{N}\{u^2(t)-\sigma^2-\gamma' z(t)\}-\bar{z}'(\tilde{\gamma}-\gamma) \\ &= N^{-1}\sum_{t=1}^{N}\{\xi(t)-\bar{z}'\{N^{-1}\sum_{s=1}^{N}(z(s)-\bar{z})(z(s)-\bar{z})'\}^{-1}(z(t)-\bar{z})\xi(t)\} \\ &= N^{-1}\sum_{t=1}^{N}\alpha_t\xi(t) ,\end{aligned} \tag{3.3.8}$$

where $\alpha_t = 1 - \bar{z}'\{N^{-1}\sum_{s=1}^{N}(z(s)-\bar{z})(z(s)-\bar{z})'\}^{-1}(z(t)-\bar{z})$.

Let $K^* = [\beta^{*\prime}, \gamma^{*\prime}, \sigma^{2*}]'$, where

$$\beta^*-\beta = V^{-1}N^{-1} \sum_{t=1}^{N} Y(t-1)u(t) ,$$

$$\gamma^*-\gamma = R^{-1}N^{-1} \sum_{t=1}^{N} (z(t)-E(z(t)))\xi(t)$$

and

$$\sigma^{2*}-\sigma^2 = N^{-1} \sum_{t=1}^{N} \alpha_t^*\xi(t) ,$$

with $R = E\{(z(t)-E[z(t)])(z(t)-E[z(t)])'\}$, assuming that $E(X^4(t)) < \infty$, and $\alpha_t^* = 1 - E[z'(t)]R^{-1}(z(t)-E[z(t)])$. Since $N^{-1} \sum_{t=1}^{N} Y(t-1)Y'(t-1)$ converges almost surely to V, it is first seen that $(\beta^*-\hat{\beta})$ converges almost surely to 0, and $N^{\frac{1}{2}}(\beta^*-\hat{\beta})$ converges in probability to zero. Next

$$\{N^{-1} \sum_{t=1}^{N} (z(t)-\bar{z})(z(t)-\bar{z})'\}(\tilde{\gamma}-\gamma)-R(\gamma^*-\gamma) = N^{-1} \sum_{t=1}^{N} (E[z(t)]-\bar{z})\xi(t) .$$

Now $E(\xi(t)|F_{t-1}) = 0$ by (3.2.2), so that $N^{-1} \sum_{t=1}^{N} \xi(t)$ converges almost surely to zero. Also, if $E(X^4(t)) < \infty$, it is easily seen that $E(\xi^2(t)) < \infty$. Thus $N^{-\frac{1}{2}} \sum_{t=1}^{N} \xi(t)$ converges in distribution to a normal distribution. But $\bar{z}$ converges almost surely to $E[z(t)]$, so that $N^{-\frac{1}{2}} \sum_{t=1}^{N} (E[z(t)]-\bar{z})\xi(t)$ converges in probability to zero. Hence, since $N^{-1} \sum_{t=1}^{N} (z(t)-\bar{z})(z(t)-\bar{z})'$ converges almost surely to R, $(\tilde{\gamma}-\gamma^*)$ converges almost surely to zero, while $N^{\frac{1}{2}}(\tilde{\gamma}-\gamma^*)$ converges in probability to zero. Finally,

$$\sigma^{2*}-\tilde{\sigma}^2 = N^{-1} \sum_{t=1}^{N} (\alpha_t^*-\alpha_t)\xi(t)$$

and

$$(\alpha_t^*-\alpha_t) = \{\bar{z}'[N^{-1} \sum_{s=1}^{N} (z(s)-\bar{z})(z(s)-\bar{z})']^{-1} - E[z'(t)]R^{-1}\}z(t)$$
$$- \bar{z}'\{N^{-1} \sum_{s=1}^{N} (z(s)-\bar{z})(z(s)-\bar{z})'\}^{-1}\bar{z} + E[z'(t)]R^{-1}E[z(t)] .$$

Now $N^{-1}\sum_{t=1}^{N} z(t)\xi(t)$ converges almost surely to zero if $E(X^4(t)) < \infty$, since $\{z(t)\xi(t)\}$ is ergodic and

$$E[z(t)\xi(t)] = E\{z(t)E[\xi(t)|F_{t-1}]\} = 0 .$$

Furthermore, $\bar{z}'[N^{-1}\sum_{s=1}^{N}(z(s)-\bar{z})(z(s)-\bar{z})']^{-1}$ converges almost surely to $E[z'(t)]R^{-1}$ and $\bar{z}'\{N^{-1}\sum_{s=1}^{N}(z(s)-\bar{z})(z(s)-\bar{z})'\}^{-1}\bar{z}$ converges almost surely to $E[z'(t)]R^{-1}E[z(t)]$. Thus it is evident that $(\sigma^{2*}-\tilde{\sigma}^2)$ converges almost surely to zero. Moreover, $N^{-\frac{1}{2}}\sum_{t=1}^{N} z(t)\xi(t)$ will converge in distribution if $E(X^8(t)) < \infty$, while $N^{-\frac{1}{2}}\sum_{t=1}^{N}\xi(t)$ will converge in distribution if $E(X^4(t)) < \infty$, both these results being obtained by an application of theorem A.1.4.

Hence $(K^*-\tilde{K})$ converges almost surely to zero if $E(X^4(t)) < \infty$ and $N^{\frac{1}{2}}(K^*-\tilde{K})$ converges in probability to zero if $E(X^8(t)) < \infty$. We thus obtain the results for K^*, which will also apply to $\tilde{K}$ and $\hat{K}$.

Let a be any $(n+1)(n+2)/2$-component vector, and let $a = [a_1',a_2',a_3]'$, where a_1, a_2 and a_3 have $(n+1)$, $n(n+1)/2$ and 1 components respectively. Then

$$a'(K^*-K) = N^{-1}\sum_{t=1}^{N}\chi_t(a) ,$$

where

$$\chi_t(a) = a_1'V^{-1}Y(t-1)u(t) + a_2'R^{-1}(z(t)-E[z(t)])\xi(t) + a_3\{1-E[z'(t)]R^{-1}(z(t)-E[z(t)])\}\xi(t) .$$

Now, since $E(u(t)|F_{t-1}) = E(\xi(t)|F_{t-1}) = 0$, we have $E(\chi_t(a)|F_{t-1}) = 0$. Also, it is easy to see that $E(\chi_t^2(a)) < \infty$ provided that $E(X^8(t)) < \infty$. Thus, since $\{\chi_t(a)\}$ is strictly stationary and ergodic, $a'(K^*-K)$ converges almost surely to zero and from theorem A.1.4, it immediately follows that $a'N^{\frac{1}{2}}(K^*-K)$ has a distribution which converges to the normal distribution with mean zero and variance $E(\chi_t^2(a))$. But $E(\chi_t^2(a))$ may be written in the form $a'\Omega a$, where Ω is symmetric and not dependent on the vector a. Thus (K^*-K) and $(\hat{K}-K)$

converge almost surely to zero, and $N^{\frac{1}{2}}(\hat{K}-K)$ has a distribution which converges to the normal distribution with mean zero and covariance matrix Ω, defined by the following. Let Ω be given by

$$(3.3.9) \qquad \begin{bmatrix} \Omega_{11} & \Omega_{12} & \Omega_{13} \\ \Omega'_{12} & \Omega_{22} & \Omega_{23} \\ \Omega'_{13} & \Omega'_{23} & \Omega_{33} \end{bmatrix}$$

where for $i,j = 1,2,3$, Ω_{ij} is an $n(i)\times n(j)$ matrix with $n(1) = n$, $n(2) = n(n+1)/2$ and $n(3) = 1$. Then the Ω_{ij}'s, $1 \le i \le j \le 3$ are obtained by evaluating the components of $E(\chi_t^2(a))$ of the form $a_i'Ma_j$. Since the evaluation is somewhat tedious, the Ω_{ij}'s are derived in appendix 3.2. #

While the eighth moment condition on the process $\{X(t)\}$ would seem unduly restrictive, it should be remembered that only a fourth moment condition is needed for a central limit theorem to exist for $\hat{\beta}$, the estimate of the most important parameter associated with (1.1.1), as the best (not necessarily linear) predictor, in the least squares sense, of $X(t)$ given $\{X(t-1),X(t-2),...\}$ is given by

$$E(X(t)|F_{t-1}) = \sum_{i=1}^{n} \beta_i X(t-i) + E(u(t)|F_{t-1}) = \sum_{i=1}^{n} \beta_i X(t-i) ,$$

since $E(u(t)|F_{t-1}) = 0$. It will also be necessary that the eighth moments of both processes $\{B(t)\}$ and $\{\varepsilon(t)\}$ exist in order that the central limit theorem exist for $\hat{\sigma}^2$ and $\hat{\gamma}$. It is hoped, nevertheless, that, since $\hat{K}$ converges almost surely to K if $E(X^4(t)) < \infty$, and, moreover, that $\hat{\beta}$ converges almost surely to β without this moment condition, that $\hat{K}$ will prove a good initial estimate of K for the iterative maximum likelihood procedure introduced in chapter 4. This will be demonstrated in chapter 5 by means of a number of simulations.

3.4 The Consistent Estimation of the Covariance Matrix of the Estimates

As we saw in §3.3, $N^{\frac{1}{2}}(\hat{K}-K)$ has a distribution which converges to the normal distribution with zero mean and covariance matrix Ω, where Ω is defined by (3.3.9). In practice of course, having estimated the coefficients of (1.1.1), an estimate of the covariance matrix Ω is required.

Let

$$\hat{u}(t) = X(t) - \hat{\beta}'Y(t-1) \ , \quad \hat{V} = N^{-1} \sum_{t=1}^{N} Y(t-1)Y'(t-1)$$

and

$$\hat{R} = N^{-1} \sum_{t=1}^{N} (z(t)-\bar{z})(z(t)-\bar{z})' \ .$$

Then from appendix 3.2 an estimate of Ω, which is easily shown to be strongly consistent, is given by

$$\hat{\Omega} = \{\hat{\Omega}_{ij}\} \ , \quad 1 \leq i \leq j < 3 \ , \tag{3.4.1}$$

where $\hat{\Omega}_{ij}$ is the estimate of Ω_{ij} and

$$\hat{\Omega}_{11} = \hat{\sigma}^2\hat{V}^{-1} + \hat{V}^{-1}N^{-1} \sum_{t=1}^{N} \{Y(t-1)Y'(t-1)\hat{\gamma}'z(t)\}\hat{V}^{-1}$$

$$\hat{\Omega}_{12} = \hat{V}^{-1}N^{-1} \sum_{t=1}^{N} \{Y(t-1)(z(t)-\bar{z})'\hat{u}^3(t)\}\hat{R}^{-1}$$

$$\hat{\Omega}_{13} = \hat{V}^{-1}N^{-1} \sum_{t=1}^{N} \{Y(t-1)[1-(z(t)-\bar{z})'\hat{R}^{-1}\bar{z}]\hat{u}^3(t)\}$$

$$\hat{\Omega}_{22} = \hat{R}^{-1}[N^{-1} \sum_{t=1}^{N} (z(t)-\bar{z})(z(t)-\bar{z})'\{\hat{u}^4(t)-(\hat{\sigma}^2+\hat{\gamma}'z(t))^2\}]\hat{R}^{-1}$$

$$\hat{\Omega}_{23} = \hat{R}^{-1}N^{-1} \sum_{t=1}^{N} \{(z(t)-\bar{z})[\hat{u}^4(t)-(\hat{\sigma}^2+\hat{\gamma}'z(t))^2]\}-\hat{\Omega}_{22}\bar{z}$$

and

$$\hat{\Omega}_{33} = N^{-1} \sum_{t=1}^{N} \{\hat{u}^4(t)-(\hat{\sigma}^2+\hat{\gamma}'z(t))^2\}$$

$$- 2N^{-1} \sum_{t=1}^{N} [\{\hat{u}^4(t)-(\hat{\sigma}^2+\hat{\gamma}'z(t))^2\}\bar{z}'\hat{R}^{-1}(z(t)-\bar{z})+\bar{z}'\hat{\Omega}_{22}\bar{z} \ .$$

APPENDIX 3.1

PROOF OF LEMMA 3.1. Suppose there exists a non-zero vector α with $\alpha'(z(t)-E[z(t)]) = 0$ almost surely for all t. Then, if the last component $\alpha_{n(n+1)/2}$ of α is non-zero, we may rearrange the equation $\alpha'(z(t)-E[z(t)]) = 0$ to obtain

$$X^2(t) = c + \sum_{i=1}^{n-1} \gamma_i X(t)X(t-i) + \sum_{1\leq i\leq j}^{n-1} \lambda_{ij} X(t-i)X(t-j) \tag{A.3.1}$$

almost surely, where $c = \alpha' E[z(t)]/\alpha_{n(n+1)/2}$ and the γ_i and λ_{ij} are other elements of the vector α divided by $-\alpha_{n(n+1)/2}$. If, however, $\alpha_{n(n+1)/2}$ is zero, then without loss of generality there will be a smallest integer $k \leq n$ such that the coefficient of $X(t)X(t-k)$ in $\alpha' E[z(t)]$ is non-zero. For if none of the coefficients of $X(t)X(t-k)$ $k = 1,\ldots,2$ were non-zero, then one of the coefficients of $X(t-i)X(t-j)$ would be non-zero for $1 \leq i \leq j$. Then we could find another vector α_1 such that $\alpha_1'(z(t-i) - E[z(t-i)]) = 0$ for $i = 1,\ldots,n$, and since $z(t)$ is strictly stationary, it follows that $\alpha_1'(z(t) - E[z(t)]) = 0$. Hence, in this case, $\alpha'(z(t) - E[z(t)]) = 0$ may be restated as

$$X(t)X(t-k) = c + \sum_{i=k+1}^{n} \gamma_i X(t)X(t-i) + \sum_{1\leq i\leq j}^{n-1} \lambda_{ij} X(t-i)X(t-j) \tag{A.3.2}$$

almost surely, for some constants c, γ_i and λ_{ij}. Now, (A.3.1), which is quadratic in $X(t)$, has two solutions given by

$$X(t) = \frac{1}{2}\left\{\sum_{i=1}^{n-1} \gamma_i X(t-i) \pm \left[\left(\sum_{i=1}^{n-1} \gamma_i X(t-i)\right)^2 + 4c + 4 \sum_{1\leq i\leq j}^{n-1} \lambda_{ij} X(t-i)X(t-j)\right]^{\frac{1}{2}}\right\} \tag{A.3.3}$$

Denoting these two solutions by $f_1(t)$ and $f_2(t)$, and letting A and B be the sets on which $X(t) = f_1(t)$ and $f_2(t)$ respectively, we have $Pr(A \cup B) = 1$. Furthermore, $f_1(t)$ and $f_2(t)$ depend only on $Y(t-1)$. But $X(t) = \beta' Y(t-1) + B(t)Y(t-1) + \varepsilon(t)$, so that $\varepsilon(t) = f_1(t) - \beta' Y(t-1) - B(t)Y(t-1)$

on A and $\varepsilon(t) = f_2(t) - \beta'Y(t-1) - B(t)Y(t-1)$ on B. However, $\varepsilon(t)$ is independent of $Y(t-1)$ and $B(t)$, and so must be constant on each of A and B, contradicting the assumption (vii).

If $X(t-k) \neq \sum_{i=k+1}^{n-1} \gamma_i X(t-i)$, then (A.3.2) has the solution

$$\text{(A.3.4)} \qquad X(t) = \left\{c + \sum_{1\leq i\leq j}^{n-1} \lambda_{ij} X(t-i)X(t-j)\right\} \left\{X(t-k) - \sum_{i=k+1}^{n-1} \gamma_i X(t-i)\right\}^{-1} .$$

Letting A be the set on which $X(t-k) = \sum_{i=k+1}^{n-1} \gamma_i X(t-i)$ and B its complement, and using the same argument as before, it is seen that $\varepsilon(t-k) = c_1$ on A and $\varepsilon(t) = c_2$ on B, where c_1 and c_2 are constants. But $\Pr(\varepsilon(t) = c_1) = \Pr(\varepsilon(t-k) = c_1) = \Pr(A) = 1-\Pr(B)$, since $\{\varepsilon(t)\}$ is strictly stationary. Hence $\varepsilon(t)$ is equal to c_1 or c_2 almost everywhere, again contradicting assumption (vii), and hence proving the lemma. #

PROOF OF LEMMA 3.2. From (3.2.4), (3.2.5), (3.3.1) and (3.3.2) it is seen that

$$\text{(3.3.3)} \qquad \hat{\gamma}-\tilde{\gamma} = \left\{\sum_{t=1}^{N} (z(t)-\bar{z})(z(t)-\bar{z})'\right\}^{-1} \left\{\sum_{t=1}^{N} (z_t-\bar{z})(\hat{u}^2(t)-u^2(t))\right\}$$

and

$$\text{(3.3.4)} \qquad \hat{\sigma}^2-\tilde{\sigma}^2 = N^{-1} \sum_{t=1}^{N} (\hat{u}^2(t)-u^2(t)) - (\hat{\gamma}-\tilde{\gamma})'\bar{z} .$$

Now, $\hat{u}^2(t) - u^2(t) = (\hat{u}(t)-u(t))(\hat{u}(t)+u(t))$ so that

$$\text{(3.3.5)} \qquad \hat{u}^2(t)-u^2(t) = \{X(t)-\hat{\beta}'Y(t-1)-X(t)+\beta'Y(t-1)\}\{X(t)-\hat{\beta}Y(t-1)+X(t)-\beta Y(t-1)\}$$
$$= \{(\beta-\hat{\beta})'Y(t-1)\}\{2u(t)+(\beta-\hat{\beta})'Y(t-1)\} .$$

Thus

$$(3.3.6) \qquad N^{-1}\sum_{t=1}^{N}(z(t)-\bar{z})(\hat{u}^2(t)-u^2(t))$$

$$= N^{-1}\sum_{t=1}^{N}(z(t)-\bar{z})\{(\beta-\hat{\beta})'Y(t-1)\}\{2u(t)+(\beta-\hat{\beta})'Y(t-1)\}$$

$$= 2N^{-1}\sum_{t=1}^{N}(z(t)-\bar{z})u(t)\{(\beta-\hat{\beta})'Y(t-1)\}+N^{-1}\sum_{t=1}^{N}(z(t)-\bar{z})\{(\beta-\hat{\beta})'Y(t-1)\}^2 .$$

Now

$$N^{-1}\sum_{t=1}^{N}(z(t)-\bar{z})u(t)\{(\beta-\hat{\beta})'Y(t-1)\}$$

$$= N^{-1}\sum_{t=1}^{N}z(t)u(t)\{(\beta-\hat{\beta})'Y(t-1)\} - \bar{z}N^{-1}\sum_{t=1}^{N}u(t)(\beta-\hat{\beta})'Y(t-1) .$$

The term $N^{-1}\sum_{t=1}^{N}z(t)u(t)\{(\beta-\hat{\beta})'Y(t-1)\}$ converges almost surely to zero since $(\beta-\hat{\beta})$ converges almost surely to zero and $N^{-1}\sum_{t=1}^{N}z(t)u(t)Y'(t-1)$ converges almost surely to zero by the ergodic theorem if $E(X^4(t)) < \infty$, since $\{z(t)Y'(t-1)u(t)\}$ is ergodic and $E(z(t)u(t)Y'(t-1)|F_{t-1}) = z(t)Y'(t-1)E(u(t)|F_{t-1}) = 0$. Moreover, $N^{-\frac{1}{2}}\sum_{t=1}^{N}z(t)u(t)\{(\beta-\hat{\beta})'Y(t-1)\}$ will converge in probability to zero under the same conditions, since $N^{\frac{1}{2}}(\beta-\hat{\beta})$ converges in distribution. Also, $\bar{z}N^{-1}\sum_{t=1}^{N}u(t)(\beta-\hat{\beta})'Y(t-1)$ converges almost surely to zero since $\bar{z}$ converges almost surely to $E(z(t))$ by the ergodic theorem and $N^{-1}\sum_{t=1}^{N}u(t)Y(t-1)$ converges almost surely to zero. Again, it is true that $\bar{z}N^{-\frac{1}{2}}\sum_{t=1}^{N}u(t)(\beta-\hat{\beta})'Y(t-1)$ converges in probability to zero.

Consider now the second term on the right hand side of (3.3.6),

$$N^{-1}\sum_{t=1}^{N}(z(t)-\bar{z})\{(\beta-\hat{\beta})'Y(t-1)\}^2$$

$$= N^{-1}\sum_{t=1}^{N}z(t)\{(\beta-\hat{\beta})'Y(t-1)\}^2 - \bar{z}N^{-1}\sum_{t=1}^{N}\{(\beta-\hat{\beta})'Y(t-1)\}^2 .$$

Since $Y(t-1)$ and $(\beta-\hat{\beta})$ are both $n\times 1$ vectors, it follows from result 2 of theorem A.1.1 that

$$[\text{vec}\{Y(t-1)Y'(t-1)\}]'\text{vec}\{(\beta-\hat{\beta})(\beta-\hat{\beta})'\} = \text{tr}\{Y(t-1)Y'(t-1)(\beta-\hat{\beta})(\beta-\hat{\beta})'\}$$

$$= \text{tr}[(\beta-\hat{\beta})'Y(t-1)]^2 = [(\beta-\hat{\beta})'Y(t-1)]^2 .$$

Hence

$$N^{-1}\sum_{t=1}^{N} z(t)\{(\beta-\hat{\beta})'Y(t-1)\}^2$$

$$= N^{-1}\sum_{t=1}^{N} z(t)[\text{vec}(Y(t-1)Y'(t-1))]'\text{vec}[(\beta-\hat{\beta})(\beta-\hat{\beta})']$$

and $N^{-1}\sum_{t=1}^{N} z(t)[\text{vec}(Y(t-1)Y'(t-1))]'$ converges almost surely by the ergodic theorem to $E\{z(t)[\text{vec}(Y(t-1)Y'(t-1))]'\}$ if $E(X^4(t)) < \infty$. Since $(\beta-\hat{\beta})$ converges almost surely to zero, and $N^{\frac{1}{2}}(\beta-\hat{\beta})$ converges in distribution, implying that $N^{\frac{1}{4}}(\beta-\hat{\beta})$ converges in probability to zero, it follows simply that $N^{-1}\sum_{t=1}^{N} z(t)\{(\beta-\hat{\beta})'Y(t-1)\}^2$ and $N^{-1}\sum_{t=1}^{N} \bar{z}\{(\beta-\hat{\beta})'Y(t-1)\}^2$ converge almost surely to zero, while $N^{-\frac{1}{2}}\sum_{t=1}^{N} z(t)\{(\beta-\hat{\beta})'Y(t-1)\}^2$ and $N^{-\frac{1}{2}}\sum_{t=1}^{N} \bar{z}\{(\beta-\hat{\beta})'Y(t-1)\}^2$ converge in probability to zero. It should be noted that we have used the results that if $\omega(N)$ is a random vector which converges in distribution, then $N^{-\frac{1}{4}}\omega(N)$ converges in probability to zero, and that if $\xi(N)$ is another random vector which converges in probability to zero, then $\xi(N)'\omega(N)$ converges in probability to zero.

Combining these results, it is easily seen from (3.3.7) that $(\hat{\gamma}-\tilde{\gamma})$ converges almost surely to zero, while $N^{\frac{1}{2}}(\hat{\gamma}-\tilde{\gamma})$ converges in probability to zero, since

$$N^{-1}\sum_{t=1}^{N}(z(t)-\bar{z})(z(t)-\bar{z})'$$

$$= N^{-1}\sum_{t=1}^{N} z(t)z'(t) - \bar{z}N^{-1}\sum_{t=1}^{N} z'(t) - N^{-1}\sum_{t=1}^{N} z(t)\bar{z} + \bar{z}\bar{z}'$$

which converges almost surely to

$$E(z(t)z'(t))-E(z(t))E(z'(t)) = E\{[z(t)-E(z(t))][z(t)-E(z(t))]'\} .$$

Because of the fact that $E((X^4(t)) < \infty)$, and by lemma 3.1, this matrix is positive definite.

Using the same arguments, it is easily seen that

$$N^{-1}\sum_{t=1}^{N}(\hat{u}^2(t)-u^2(t))$$

$$= N^{-1}\sum_{t=1}^{N}(\beta-\hat{\beta})'Y(t-1)u(t) + N^{-1}\sum_{t=1}^{N}\{(\beta-\hat{\beta})'Y(t-1)\}^2$$

converges almost surely to zero, while $N^{-\frac{1}{2}}\sum_{t=1}^{N}(\hat{u}^2(t)-u^2(t))$ converges in probability to zero, even without the assumption that $E(X^4(t)) < \infty$. Thus, from (3.3.4), $(\hat{\sigma}^2-\tilde{\sigma}^2)$ converges almost surely to zero, while $N^{\frac{1}{2}}(\hat{\sigma}^2-\tilde{\sigma}^2)$ converges in probability to zero, since $\bar{z}$ converges almost surely to $E(z(t))$. #

APPENDIX 3.2

The Derivation of the Covariance Matrix Ω

The submatrices Ω_{ij}, $1 \leq i \leq j \leq 3$, of Ω defined in (3.3.9) are found by evaluating those components of $E(\chi_t^2(a))$ of the form $a_i'Ma_j$. The submatrix Ω_{11} is given by

$$\Omega_{11} = \sigma^2 V^{-1} + V^{-1}E\{Y(t-1)Y'(t-1)\gamma' z(t)\}V^{-1} ,$$

which is derived in theorem 3.1, while

$$\begin{aligned}\Omega_{12} &= E\{V^{-1}Y(t-1)(z(t)-E[z(t)])'R^{-1}\xi(t)u(t)\} \\ &= E\{V^{-1}Y(t-1)(z(t)-E[z(t)])'R^{-1}E(\xi(t)u(t)|F_{t-1})\} .\end{aligned}$$

Now,

$$\begin{aligned}E(\xi(t)u(t)|F_{t-1}) &= E\{(u^3(t)-[\sigma^2+\gamma' z(t)]u(t))|F_{t-1}\} \\ &= E\{u^3(t)|F_{t-1}\}-[\sigma^2+\gamma' z(t)]E\{u(t)|F_{t-1}\} \\ &= E\{(\varepsilon(t)+B(t)Y(t-1))^2|F_{t-1}\} \\ &= E\{\varepsilon^3(t)\}+E\{(B(t)Y(t-1))^3|F_{t-1}\} ,\end{aligned}$$

since $E(u(t)|F_{t-1}) = 0$, $E(\varepsilon(t)|F_{t-1}) = E(\varepsilon(t)) = 0$ and $E(B(t)Y(t-1)|F_{t-1}) = E(B(t)|F_{t-1})Y(t-1) = 0$. If the distributions of $\varepsilon(t)$ and $B(t)$ are known, the above term is reducible. For example, if $\{\varepsilon(t)\}$ and $\{B(t)\}$ are normally distributed, then conditional on F_{t-1}, $u(t)$ is normal with mean zero. Hence $E(u^3(t)|F_{t-1}) = 0$. Whatever the case, the above shows that $E(\xi(t)u(t)|F_{t-1}) = E(u^3(t)|F_{t-1})$, so that

$$\Omega_{12} = V^{-1}E\{Y(t-1)(z(t)-E[z(t)])'u^3(t)\}R^{-1} ,$$

and

$$\begin{aligned}\Omega_{13} &= E\{V^{-1}Y(t-1)(1-(z(t)-E[z(t)])'R^{-1}E[z(t)])\xi(t)u(t)\} \\ &= V^{-1}E\{Y(t-1)(1-(z(t)-E[z(t)])'R^{-1}E[z(t)])u^3(t)\} .\end{aligned}$$

Also,

$$\Omega_{22} = R^{-1}E\{(z(t)-E[z(t)])\xi^2(t)(z(t)-E[z(t)])'\}R^{-1} .$$

Now

$$E(\xi^2(t)|F_{t-1}) = E\{(u^2(t)-\sigma^2-\gamma' z(t))^2|F_{t-1}\}$$
$$= E\{u^4(t)|F_{t-1}\}-2(\sigma^2+\gamma' z(t))E(u^2(t)|F_{t-1})+(\sigma^2+\gamma' z(t))^2$$
$$= E(u^4(t)|F_{t-1})-(\sigma^2+\gamma' z(t))^2 \ .$$

Again this may be simplified if more is known about the distributions of $\{\varepsilon(t)\}$ and $\{B(t)\}$. For example, if $\{\varepsilon(t)\}$ and $\{B(t)\}$ are normally distributed, then $E(u^4(t)|F_{t-1}) = 3\{E(u^2(t)|F_{t-1})\}^2 = 3(\sigma^2+\gamma' z(t))^2$ and $E(\xi^2(t)|F_{t-1}) = 2(\sigma^2+\gamma' z(t))$. In any case, however

$$\Omega_{22} = R^{-1}E\{(z(t)-E[z(t)])(z(t)-E[z(t)])'(u^4(t)-(\sigma^2+\gamma' z(t))^2)\}R^{-1} \ ,$$

$$\Omega_{23} = E\{R^{-1}(z(t)-E[z(t)])(1-(z(t)-E[z(t)])'R^{-1}E[z(t)])\xi^2(t)\}$$
$$= R^{-1}E\{(z(t)-E[z(t)])(u^4(t)-(\sigma^2+\gamma' z(t))^2)\}$$
$$-R^{-1}E\{(z(t)-E[z(t)])(z(t)-E[z(t)])'(u^4(t)-(\sigma^2+\gamma' z(t))^2)\}R^{-1}E[z(t)]$$
$$= R^{-1}E\{(z(t)-E[z(t)])(u^4(t)-(\sigma^2+\gamma' z(t))^2)\}-\Omega_{22}E[z(t)]$$

and

$$\Omega_{33} = E\{(1-E[z'(t)]R^{-1}(z(t)-E[z(t)]))^2\xi^2(t)\}$$
$$= E\{u^4(t)-(\sigma^2+\gamma' z(t))^2\}$$
$$-2E\{(u^4(t)-(\sigma^2+\gamma' z(t))^2)E[z'(t)]R^{-1}(z(t)-E[z(t)])\}+E[z'(t)]\Omega_{22}E[z(t)]$$

CHAPTER 4

MAXIMUM LIKELIHOOD ESTIMATION OF SCALAR MODELS

4.1 Introduction

An estimation procedure is introduced in this chapter which is based on maximizing the likelihood function constructed as though the processes $\{\varepsilon(t)\}$ and $\{B(t)\}$ were sequences of normally distributed random variables. We shall refer to the estimates obtained in this way as maximum likelihood estimates even though it will be shown that these estimates will be strongly consistent and satisfy a central limit theorem if the processes $\{\varepsilon(t)\}$ and $\{B(t)\}$ are not sequences of normally distributed random variables. It will be seen that the likelihood function to be optimized is non-linear in the parameters (to be estimated) and so it is necessary to use an iterative procedure to obtain the estimates. As was mentioned earlier, it is desirable to commence such an iterative procedure as close to the global optimum as possible in order to reduce the possibility of converging to a local optimum. For this reason we choose strongly consistent estimates of the parameters to commence the iterative procedure, such estimates having been derived in the previous chapter.

The moment conditions required to prove the asymptotic properties of the least squares estimates considered in chapter 3 will not be required here. Indeed, in order that a central limit theorem exist for the maximum likelihood estimates, conditions (i)-(vii) will be required, along with the condition

(viii) $\qquad E(\varepsilon^4(t)) < \infty$ and $E(B_i^4(t)) < \infty, \quad i = 1,\ldots,n.$

The estimates $\hat{\sigma}^2$ and $\hat{\gamma}$ of σ^2 and γ respectively will be such that $\hat{\sigma}^2 \geq 0$ and the matrix $\hat{\Sigma}$, where $\hat{\gamma} = \text{vech}\, \hat{\Sigma}$, is non-negative definite. Thus a central limit theorem for $\hat{\sigma}^2$ and $\hat{\gamma}$ would involve certain complications if either $\sigma^2 = 0$ or Σ had a zero eigenvalue. To avoid these complications we assume the following condition

(ix) $\sigma^2 \geq \delta_1 > 0$ while the smallest eigenvalue of Σ is bounded below by δ_2, where δ_1 and δ_2 may be taken as small as required

We shall also need to have the second moments of $\{X(t)\}$ bounded. In light of the necessary and sufficient conditions for condition (vi) to hold, discussed at the end of §3.1, we replace (vi) with the stronger assumption

(vi)' The eigenvalues of M have moduli bounded above by $(1-\delta_3) < 1$, while $(\text{vec } \Sigma)' \text{ vec } W$ is bounded above by $(1-\delta_4) < 1$, where δ_3 and δ_4 are both arbitrarily small, and W is defined in §3.1.

By corollary 2.3.2, the covariance matrix $V = E(Y(t)Y'(t))$ is given by $V = \sigma^2\{1-(\text{vec } \Sigma)'\text{vec } W\}^{-1}W$. It will be seen in the proof of lemma 4.1 that if the eigenvalues of the matrix M are bounded above by $(1-\delta_3) < 1$, then the eigenvalues of W will be bounded above. Thus, since $(1-(\text{vec } \Sigma)'\text{vec } W)$ is bounded below by $\delta_4 > 0$, the matrix V is bounded above. This fact will be needed when we prove the strong consistency of the maximum likelihood procedure, as well as in the proof of the central limit theorem.

Henceforth we shall refer to condition (vi)' as condition (vi), for the sake of the uniformity of notation.

4.2 The Maximum Likelihood Procedure

Given a sample $\{X(1),\ldots,X(N)\}$ from a time series $\{X(t)\}$ which is strictly stationary, F_t-measurable and satisfies (1.1.1) under conditions (i)-(ix), we shall derive the likelihood function conditional on preperiod values $\{X(1-n),\ldots,X(0)\}$, as though we were assuming the joint normality of $\{\varepsilon(t)\}$ and $\{B(t)\}$. Let $f_s(X(t),\ldots,X(t-s+1)|A_{t-s})$ denote the density of $X(t),\ldots,X(t-s+1)$ given an event A_{t-s} in the σ-field F_{t-s}. Then from the structure of (1.1.1), we have

$$(4.2.1)\qquad E(X(t)|Y(t-1)) = E\{[\sum_{i=1}^{n}(\beta_i+B_i(t))X(t-i)+\varepsilon(t)]|Y(t-1)\}$$
$$= \beta'Y(t-1)$$

and

$$(4.2.2)\qquad \text{Var}\{X(t)|Y(t-1)\} = E\{[B(t)Y(t-1)+\varepsilon(t)]^2|Y(t-1)\}$$
$$= E\{[Y'(t-1)B'(t)B(t)Y(t-1)+2\varepsilon(t)B(t)Y(t-1)$$
$$+\varepsilon^2(t)]|Y(t-1)\}$$
$$= Y'(t-1)\Sigma Y(t-1) + \sigma^2$$
$$= \gamma'z(t) + \sigma^2,$$

where $z(t) = K_n\text{vec}\{Y(t-1)Y'(t-1)\}$ as in chapter 3, and $\gamma = \text{vech}\ \Sigma$. Hence

$$(4.2.3)\qquad f_N\{X(1),\dots,X(N)|X(0),\dots,X(1-n)\}$$
$$= \prod_{t=1}^{N} f_1\{X(t)|X(t-1),\dots,X(t-n)\}$$
$$= (2\pi)^{-N/2}\sum_{t=1}^{N}\{(\sigma^2+\gamma'z(t))^{-\frac{1}{2}}\exp[-\tfrac{1}{2}\frac{(X(t)-\beta'Y(t-1))^2}{\sigma^2+\gamma'z(t)}]\}$$
$$= L_N(\beta,\gamma,\sigma^2)$$

which is the likelihood function conditional on $\{X(0),\dots,X(1-n)\}$. It will prove more convenient to consider, instead of the maximization of $L_N(\beta,\gamma,\sigma^2)$, the minimization of the function

$$(4.2.4)\qquad \tilde{\ell}_N(\beta,\gamma,\sigma^2) = -2/N\ \ln\{L_N(\beta,\gamma,\sigma^2)\} - \ln(2\pi)$$
$$= N^{-1}\sum_{t=1}^{N}\ln(\sigma^2+\gamma'z(t)) + N^{-1}\sum_{t=1}^{N}\frac{(X(t)-\beta'Y(t-1))^2}{\sigma^2+\gamma'z(t)}.$$

The function $\tilde{\ell}_N(\beta,\gamma,\sigma^2)$ is non-linear in σ^2 and γ, and there is no closed form expression for the estimates $\hat{\beta}_N$, $\hat{\gamma}_N$ and $\hat{\sigma}^2_N$ of β, γ and σ^2, respectively, which minimize $\tilde{\ell}_N$. Nevertheless, by letting $r = \gamma/\sigma^2$ we may equivalently minimize a function of r alone, by concentrating out the parameters β and σ^2. For, letting $\bar{\ell}_N(\beta,r,\sigma^2) = \tilde{\ell}_N(\beta,\gamma,\sigma^2)$, where $r = \gamma/\sigma^2$, we have

(4.2.5) $\bar{\ell}_N(\beta,r,\sigma^2)$

$$= \ln \sigma^2 + N^{-1} \sum_{t=1}^{N} \ln(1+r'z(t)) + \sigma^{-2}N^{-1} \sum_{t=1}^{N} \frac{(X(t)-\beta'Y(t-1))^2}{1+r'z(t)} .$$

But

$$\frac{\partial}{\partial\beta} \bar{\ell}_N(\beta,r,\sigma^2) = -2\sigma^{-2}N^{-1} \sum_{t=1}^{N} \frac{(X(t)-\beta'Y(t-1))Y(t-1)}{1+r'z(t)}$$

and

$$\frac{\partial}{\partial\sigma^2} \bar{\ell}_N(\beta,r,\sigma^2) = \sigma^{-2} - \sigma^{-4}N^{-1} \sum_{t=1}^{N} \frac{(X(t)-\beta'Y(t-1))^2}{1+r'z(t)} .$$

Now $\frac{\partial}{\partial\beta} \bar{\ell}_N(\beta,r,\sigma^2) = 0$ only when

$$N^{-1} \sum_{t=1}^{N} \frac{X(t)Y(t-1)}{1+r'z(t)} = N^{-1} \sum_{t=1}^{N} \frac{Y(t-1)Y'(t-1)}{1+r'z(t)} \beta,$$

that is, when

$$\beta = \beta_N(r) = \left\{N^{-1} \sum_{t=1}^{N} \frac{Y(t-1)Y'(t-1)}{1+r'z(t)}\right\}^{-1} N^{-1} \sum_{t=1}^{N} \frac{X(t)Y(t-1)}{1+r'z(t)} .$$

Also, $\frac{\partial}{\partial(\beta',\sigma^2)'} \bar{\ell}_N(\beta,r,\sigma^2) = 0$ only when

$$\sigma^2 = \sigma_N^2(r) = N^{-1} \sum_{t=1}^{N} \frac{(X(t)-\beta_N'(r)Y(t-1))^2}{1+r'z(t)} .$$

Thus, the maximum likelihood estimates $\hat{\beta}_N$, $\hat{\gamma}_N$ and $\hat{\sigma}_N^2$ may be obtained by calculating $\hat{r}_N$, where $\hat{r}_N$ minimizes the function $\ell_N^*(r) = \ln\{\sigma_N^2(r)\} + N^{-1} \sum_{t=1}^{N} \ln(1+r'z(t))$ and $\sigma_N^2(r)$ is given above, the estimates $\hat{\beta}_N$, $\hat{\gamma}_N$ and $\hat{\sigma}_N^2$ being given by $\hat{\sigma}_N^2 = \sigma_N^2(\hat{r}_N)$, $\hat{\beta}_N = \beta_N(\hat{r}_n)$ and $\hat{\gamma}_N = \hat{\sigma}_N^2\hat{r}_N$. It is noted in passing that for N large enough, the matrix $N^{-1} \sum_{t=1}^{N} \frac{Y(t-1)Y'(t-1)}{1+r'z(t)}$ will be invertible almost everywhere, since it is obviously non-negative definite, and if it were <u>not</u> positive definite, there would exist a non-zero n-component vector α such that $\alpha'Y(t-1) = 0$ almost everywhere. This is precluded, however, because of assumption (vi).

The procedure above would be useful if one were using an optimization algorithm not requiring the first and second derivatives of the function $\ell_N^*(r)$, for these derivatives are complicated, and some loss of accuracy may be involved in their computation. Moreover, we shall be interested in obtaining a central limit theorem for $\hat{\beta}_N$, $\hat{\gamma}_N$ and $\hat{\sigma}_N^2$, and so a theoretical examination of the estimate $\hat{r}_N$ would introduce complications in obtaining this central limit theorem since $\hat{\gamma}_N = \hat{\sigma}_N^2\hat{r}_N$ and $\hat{\beta}_N = \beta_N(\hat{r}_N)$. Consequently it is better to minimize the function of β and r,

(4.2.6) $$\ell_N(\beta,r) = \inf_{\sigma^2} \tilde{\ell}_N(\beta,r,\sigma^2) - 1$$

$$= N^{-1}\sum_{t=1}^{N} \ln(1+r'z(t)) + \ln\left\{N^{-1}\sum_{t=1}^{N}\frac{(X(t)-\beta'Y(t-1))^2}{1+r'z(t)}\right\}$$

the latter expression following directly from (4.2.5). The maximum likelihood estimates $\hat{\beta}_N$, $\hat{\gamma}_N$ and $\hat{\sigma}_N^2$ are defined by

(4.2.7) $$\hat{\ell}_N(\hat{\beta}_N,\hat{r}_N) = \inf_{(\beta',r')'\in\Theta} \ell_N(\beta,r) ,$$

(4.2.8) $$\hat{\sigma}_N^2 = N^{-1}\sum_{t=1}^{N}\frac{(X(t)-\hat{\beta}_N Y(t-1))^2}{1+\hat{r}_N' z(t)}$$

and

(4.2.9) $$\hat{\gamma}_N = \hat{\sigma}_N^2\hat{r}_N.$$

The set Θ is defined in 4.3, where the strong consistency of $\hat{\beta}_N$, $\hat{\gamma}_N$ and $\hat{\sigma}_N^2$ is also shown by means of an examination of $\ell_N(\beta,r)$. For reasons of convenience, however, the central limit theorem will be proved from an examination of the unconcentrated log-likelihood function $\tilde{\ell}_N(\beta,\gamma,\sigma^2)$. Since the same estimates are obtained by the minimization of either function, the differences in approach will prove of no importance.

4.3 The Strong Consistency of the Estimates

The set Θ over which $\ell_N(\beta,r)$ is to be minimized depends on three positive numbers: δ_3, defined in (vi), δ_5 and δ_6, where δ_5 may be taken as arbitrarily small. Θ is defined as the set of all vectors $[\beta',r']'$, with β and r having n and $n(n+1)/2$ components respectively, satisfying the following conditions

(ci) β is such that all the eigenvalues of the matrix M, defined in §2.4, have moduli less than or equal to $(1-\delta_3)$;

(cii) Letting R be the square symmetric matrix for which $r = \text{vech } R$, then R has strictly positive eigenvalues, all of which are larger than or equal to δ_5;

(ciii) $(\text{vec } R)'w \leq \delta_6$, where w is the last column of $(I-M \otimes M)^{-1}$.

Suppose now that $\theta_0 = [\beta_0',r_0']'$ and that $\{X(t)\}$ is a strictly stationary, F_t-measurable solution to (1.1.1) satisfying conditions (i)-(ix) and for which $\beta = \beta_0$, $\gamma = \gamma_0$, $\sigma^2 = \sigma_0^2$ and $r = r_0 = \gamma_0/\sigma_0^2$.

The proof of the strong consistency of the maximum likelihood estimates will require that Θ be compact in order that several results from real analysis may be used. In particular we shall need to know that any function continuous on Θ achieves its supremum and infimum on Θ, and that equicontinuity and uniform convergence on Θ are equivalent. In view of this we now state lemma 4.1.

LEMMA 4.1. <u>The set</u> Θ <u>is a compact subset of</u> $\mathbb{R}^{n(n+3)/2}$ <u>for suitable</u> δ_3, δ_5 <u>and</u> δ_6, <u>and</u> δ_3, δ_5 <u>and</u> δ_6 <u>may be chosen so that</u> $\theta_0 \in \text{int}(\Theta)$.

Proof. See appendix 4.1.

The following lemma will prove useful in determining the behaviour of terms such as $N^{-1}\sum_{t=1}^{N}\frac{Y(t-1)Y'(t-1)}{1+r'z(t)}$ which will constantly appear in the proofs of strong consistency and the central limit theorem.

LEMMA 4.2. *Let* Ω *be a* $p\times p$ *symmetric positive definite matrix with smallest eigenvalue* λ_1. *Then the matrix* $zz'/(1+z'\Omega z)$ *is bounded above and below element by element for all* $z \in \mathbb{R}^p$.

Proof. See appendix 4.1.

The following theorem provides one of the main results required in the proof of the strong consistency of $\hat{\beta}_N$, $\hat{\gamma}_N$, $\hat{\sigma}^2_N$.

THEOREM 4.1. *Let* $\{X(t)\}$ *be strictly stationary,* F_t*-measurable and satisfy* (1.1.1) *with* $\beta = \beta_0$, $\sigma^2 = \sigma^2_0$ *and* $\gamma = \gamma_0$ *under conditions* (i)-(vii) *and* (ix), *and let* $\theta_0 = [\beta_0', r_0']'$, *where* $r_0 = \gamma_0/\sigma^2_0$. *Then* $\lim_{N\to\infty} \ell_N(\beta,r)$ *exists almost surely for all* $[\beta', r'] \in \Theta$, *and the limit* $\ell(\beta,r)$ *is uniquely minimized over* Θ *at* $[\beta',r']' = \theta_0$, *provided that* $\theta_0 \in \text{int}(\Theta)$.

Proof. Since $0 \leq \ln(1+r'z(t)) \leq r'z(t)$, and $E[z(t)]$ exists by (vi), $N^{-1}\sum_{t=1}^{N} \ln(1+r'z(t))$ converges almost surely to $E[\ln(1+r'z(t))]$ by the ergodic theorem. Also, since

$$0 < N^{-1}\sum_{t=1}^{N} \frac{(X(t)-\beta'Y(t-1))^2}{1+r'z(t)} \leq N^{-1}\sum_{t=1}^{N} (X(t)-\beta'Y(t-1))^2$$

and this latter term has finite expectation and converges by the ergodic theorem, we have

$$N^{-1}\sum_{t=1}^{N} \frac{(X(t)-\beta'Y(t-1))^2}{1+r'z(t)} \xrightarrow{a.s.} E\left[\frac{(X(t)-\beta'Y(t-1))^2}{1+r'z(t)}\right] .$$

Moreover, the right-hand side of the above is strictly greater than zero, for otherwise we would have $X(t) = \beta'Y(t-1)$ almost surely, which is precluded by condition (vi). Hence from (4.2.6) $\ell_N(\beta,r)$ converges almost surely to $\ell(\beta,r) = E[\ln(1+r'z(t))] + \ln\left\{E\left[\frac{(X(t)-\beta'Y(t-1))^2}{1+r'z(t)}\right]\right\}$. Now

$$E[(X(t)-\beta'Y(t-1))^2/(1+r'z(t))]$$

$$= E\{[(X(t)-\beta_0'Y(t-1))^2+2(X(t)-\beta_0'Y(t-1))(\beta_0-\beta)'Y(t-1) + ((\beta_0-\beta)'Y(t-1))^2]/(1+r'z(t))\}$$

$$= E\{[(B(t)Y(t-1)+\varepsilon(t))^2+((\beta_0-\beta)'Y(t-1))^2]/(1+r'z(t))\}$$

$$= E\{[E(B(t)Y(t-1)+\varepsilon(t))^2|F_{t-1}]/(1+r'z(t))\} + E\{((\beta_0-\beta)'Y(t-1))^2/(1+r'z(t))\}$$

$$\geq E\{(\sigma_0^2+\gamma_0'z(t))/(1+r'z(t))\} = \sigma_0^2E\{(1+r_0'z(t))/(1+r'z(t))\} ,$$

since

$$E\{(X(t)-\beta_0'Y(t-1))|F_{t-1}\} = 0 ,$$

$$E\{(B(t)Y(t-1)+\varepsilon(t))^2|F_{t-1}\} = \sigma_0^2 + \gamma_0'z(t) ,$$

and $\gamma_0 = \sigma_0^2 r_0$. Moreover, equality will hold in the above only when $(\beta_0-\beta)'Y(t-1) = 0$ almost surely, that is, when $\beta = \beta_0$. Thus

$$\inf_\beta \ell(\beta,r) = \ell(\beta_0,r) = \ln(\sigma_0^2) + \ln\{E[(1+r_0'z(t))/(1+r'z(t))]\} + E[\ln(1+r'z(t))]$$

and $\ell(\beta,r) = \inf_\beta \ell(\beta,r)$ only at $\beta = \beta_0$. Now, if X is any positive random variable with expectation 1, then $E(\ln(X)) \leq \ln[E(X)] = 0$, by Jensen's inequality, with equality only when X = 1 almost surely. Letting $X = c^{-1} \dfrac{1+r_0'z(t)}{1+r'z(t)}$, where $c = E\left\{\dfrac{1+r_0'z(t)}{1+r'z(t)}\right\}$, it is seen that

$$E\left\{\ln\left(\frac{1+r'z(t)}{1+r_0'z(t)}\right)\right\} \geq -\ln\left\{E\left(\frac{1+r_0'z(t)}{1+r'z(t)}\right)\right\}$$

with equality only when $(1+r_0'z(t)) = c(1+r'z(t))$ almost surely, that is when $(r_0-cr)'z(t) = (c-1)$ almost surely. However, by lemma 3.1, this occurs only when $r_0 = cr$ and $c = 1$, that is, when $r = r_0$. Hence $\ell(\beta,r)$ is uniquely minimized at $\beta = \beta_0$ and $r = r_0$. #

COROLLARY 4.1.1. <u>$\lim_{N\to\infty} \tilde{\ell}_N(\beta,\gamma,\sigma^2)$ exists almost surely and is minimized uniquely at</u> $\beta = \beta_0$, $\gamma = \gamma_0$ <u>and</u> $\sigma^2 = \sigma_0^2$.

<u>Proof.</u> From theorem 4.1 and the definition of $\ell_N(\beta,r)$, $\lim_{N\to\infty} \tilde{\ell}_N(\beta,\gamma,\sigma^2)$ is seen to exist almost everywhere, and to be uniquely minimized at $\beta = \beta_0$, $\sigma^2 = \sigma^{2*} = E[(X(t)-\beta_0'Y(t-1))^2/(1+r_0'z(t))]$ and $\gamma = r_0\sigma^{2*}$. But

$$\sigma^{2*} = E\{E[(X(t)-\beta_0'Y(t-1))^2|F_{t-1}]/(1+r_0'z(t))\}$$

$$= E[(\sigma_0^2+\gamma_0'z(t))/(1+r_0'z(t))] = \sigma_0^2,$$

and so $\lim_{N\to\infty} \tilde{\ell}_N(\beta,\gamma,\sigma^2)$ is uniquely minimized at $\beta = \beta_0$, $\sigma^2 = \sigma_0^2$ and $\gamma = \gamma_0$. #

We are now in a position to prove the strong consistency of the procedure.

THEOREM 4.2. <u>Let</u> $\ell_N(\beta,r)$ <u>be minimized over</u> Θ <u>at</u> $\beta = \hat{\beta}_N$, $r = \hat{r}_N$, <u>and let</u> $\hat{\theta}_N = (\hat{\beta}_N',\hat{r}_N')'$. <u>Then</u> $\hat{\theta}_N$ <u>converges almost surely to</u> θ_0 <u>provided that</u> $\theta_0 \in \text{int}(\Theta)$.

<u>Proof.</u> We first show that $\{\ell_N(\beta,r)\}$ converges uniformly almost surely to $\ell(\beta,r)$ on Θ. Since Θ is compact, we need only show that $\{\ell_N(\beta,r)\}$ is equicontinuous almost surely or, letting $\theta = [\beta',r']'$, that given $\varepsilon > 0$, there exists an integer N and a positive number δ, both depending on ε, such that $|\ell_N(\theta_1) - \ell_N(\theta_2)| < \varepsilon$ almost surely for $N > N^*$ whenever $\|\theta_1-\theta_2\| < \delta$.

Now since $\ell_N(\theta)$ is differentiable on Θ, we have for each $\theta_1,\theta_2 \in \Theta$ by the mean value theorem

$$\ell_N(\theta_1)-\ell_N(\theta_2) = (\theta_1-\theta_2)' \frac{\partial}{\partial\theta} \ell_N(\theta_{12}^*),$$

where $\theta_{12}^* = \lambda\theta_1 + (1-\lambda)\theta_2$ for some $\lambda \in (0,1)$.

Let $\Theta^* = f(\Theta,\Theta,[0,1])$, where $f\colon \mathbb{R}^{n(n+3)/2} \times \mathbb{R}^{n(n+3)/2} \times \mathbb{R} \to \mathbb{R}^{n(n+3)/2}$ is the continuous function defined by $f(\theta_1,\theta_2,\lambda) = \lambda\theta_1+(1-\lambda)\theta_2$. Then Θ^* is compact since $\Theta \times \Theta \times [0,1]$ is compact; and, since $|(\theta_1-\theta_2)' \frac{\partial}{\partial\theta} \ell_N(\theta_{12}^*)|^2 \le \|\theta_1-\theta_2\|^2 \|\frac{\partial}{\partial\theta} \ell_N(\theta_{12}^*)\|^2$ it will follow that $\{\ell_N(\theta)\}$

is equicontinuous if $\lim_{N\to\infty} \sup_{\theta\in\Theta^*} \| \frac{\partial}{\partial\theta} \ell_N(\theta) \|$ is finite almost surely. The vector $\frac{\partial}{\partial\theta} \ell_N(\theta)$ is obtained from

$$\frac{\partial}{\partial\beta} \ell_N(\theta) = -2(\sigma_N^2(\theta))^{-1} N^{-1} \sum_{t=1}^{N} \frac{(X(t)-\beta' Y(t-1))Y(t-1)}{1+r'z(t)}$$

and

$$\frac{\partial}{\partial r} \ell_N(\theta) = N^{-1} \sum_{t=1}^{N} \frac{z(t)}{1+r'z(t)} - (\sigma_N^2(\theta))^{-1} N^{-1} \sum_{t=1}^{N} \frac{(X(t)-\beta' Y(t-1))^2 z(t)}{1+r'z(t)}$$

where

$$\sigma_N^2(\theta) = N^{-1} \sum_{t=1}^{N} \frac{(X(t)-\beta' Y(t-1))^2}{1+r'z(t)} .$$

Now, if $[\beta',r']' \in \Theta^*$, then, letting R be the $n \times n$ symmetric matrix such that $r = \text{vech } R$, it can be seen that $R = \lambda\Omega_1 + (1-\lambda)\Omega_2$ for some $\lambda \in [0,1]$ where the eigenvalues of Ω_1 and Ω_2 are bounded below by $\delta_5 > 0$. Thus the smallest eigenvalue of R is bounded below by δ_5, since, for any n-component vector z, $\frac{z'Rz}{z'z} = \lambda \frac{z'\Omega_1 z}{z'z} + (1-\lambda) \frac{z'\Omega_2 z}{z'z} \geq [\lambda+(1-\lambda)]\delta_5 = \delta_5$. That

$\lim_{N\to\infty} \sup_{\theta\in\Theta^*} \| \frac{\partial}{\partial\theta} \ell_N(\theta) \|$ is finite almost surely may now be proved in a straightforward manner. For example,

$$\inf_{\theta\in\Theta^*} \sigma_N^2(\theta) \geq \inf_{\theta\in\Theta^*} N^{-1} \sum_{t=1}^{N} \frac{(X(t)-\beta' Y(t-1))^2}{1+k(z'(t)z(t))^{\frac{1}{2}}}$$

where

$$k = \sup_{\theta\in\Theta^*} (r'r)^{\frac{1}{2}} ,$$

which exists since Θ^* is bounded.

Hence

$$\inf_{\theta\in\Theta^*} \sigma_N^2(\theta) \geq N^{-1} \sum_{t=1}^{N} \frac{(X(t)-\beta_N^{*\prime} Y(t-1))^2}{1+k(z'(t)z(t))^{\frac{1}{2}}}$$

where

$$\beta_N^* = \left[N^{-1} \sum_{t=1}^{N} \frac{Y(t-1)Y'(t-1)}{1+k(z'(t)z(t))^{\frac{1}{2}}}\right]^{-1} \left[N^{-1} \sum_{t=1}^{N} \frac{Y(t-1)X(t)}{1+k(z'(t)z(t))^{\frac{1}{2}}}\right].$$

However, by the ergodic theorem, β_N^* converges almost surely to β_0, and so

$$N^{-1}\sum_{t=1}^{N}\frac{(X(t)-\beta_N^{*\prime}Y(t-1))^2}{1+k(z'(t)z(t))^{\frac{1}{2}}}\xrightarrow{a.s.}E\left(\frac{(X(t)-\beta_0'Y(t-1))^2}{1+k(z'(t)z(t))^{\frac{1}{2}}}\right)$$

which is strictly greater than zero since it is not possible that $X(t)-\beta_0'Y(t-1)$ equal zero almost surely. Consequently

$$\liminf_{N\to\infty}\ \inf_{\theta\in\Theta^*}\sigma_N^2(\theta) > 0 .$$

The bounds for the other terms are obtained simply using repeated applications of lemma 4.2 and the Cauchy-Schwartz inequality.

A modification of an argument of Jennrich (1969) may now be used to show that $\hat{\theta}_N$ (any value of $\theta \in \Theta$ which minimizes $\ell_N(\theta)$) converges almost surely to θ_0. (That $\hat{\theta}_N$ exists is obvious since Θ is compact). Since $\{\ell_N(\theta)\}$ converges uniformly, for any $\varepsilon > 0$, there exists an integer N^* depending on ε such that

$$|\ell_N(\hat{\theta}_N) - \ell(\hat{\theta}_N)| < \varepsilon/2$$

and

$$|\ell_N(\theta_0) - \ell(\theta_0)| < \varepsilon/2$$

almost surely whenever $N > N^*$. Thus, since $\ell_N(\hat{\theta}_N) \le \ell_N(\theta_0)$ and $\ell(\hat{\theta}_N) \ge \ell(\theta_0)$ it follows that

$$\begin{aligned}0 \le \ell(\hat{\theta}_N)-\ell(\theta_0) &= \{\ell(\hat{\theta}_N)-\ell_N(\hat{\theta}_N)\}+\{\ell_N(\hat{\theta}_N)-\ell_N(\theta_0)\}+\{\ell_N(\theta_0)-\ell(\theta_0)\}\\ &\le \varepsilon/2 + 0 + \varepsilon/2 = \varepsilon,\end{aligned}$$

almost surely when $N > N^*$. Hence $\{\ell(\hat{\theta}_N)\}$ converges almost surely to $\ell(\theta_0)$.

Now, suppose that $\hat{\theta}_N$ does not converge almost surely to θ_0. Then it is possible to find a positive δ and an infinite subsequence $\{\hat{\theta}_{N_j}\}$ of $\{\hat{\theta}_N\}$ for which $|\hat{\theta}_{N_j}-\theta_0| \ge \delta$ for all j, on a set of positive probability. Since Θ is compact, there is an infinite subsequence $\{\hat{\theta}_{N_j'}\}$ of $\{\hat{\theta}_{N_j}\}$ which converges to, say, θ^*, where $|\theta^*-\theta_0| \ge \delta$. Thus, since $\ell(\theta)$ is continuous, it follows that

$$\lim_{N_j'\to\infty} \ell(\hat{\theta}_{N_j'}) = \ell(\theta^*) \neq \ell(\theta_0),$$

since θ_0 is the unique minimizer of $\ell(\theta)$ by theorem 4.1. However, $\lim_{N_j'\to\infty} \ell(\hat{\theta}_{N_j'}) = \lim_{N\to\infty} \ell(\hat{\theta}_N) = \ell(\theta_0)$, since $\ell(\hat{\theta}_N)$ converges almost surely to $\ell(\theta_0)$. This contradiction implies that $\hat{\theta}_N$ converges almost surely to θ_0. #

COROLLARY 4.2.1. $\hat{\gamma}_N$ and $\hat{\sigma}_N^2$ converge almost surely to γ_0 and σ_0^2 respectively.

Proof. Theorem 4.2 shows that $\hat{r}_N$ and $\hat{\beta}_N$ converge almost surely to r_0 and β_0 respectively. Now $\hat{\sigma}_N^2 = N^{-1} \sum_{t=1}^{N} \frac{(X(t)-\hat{\beta}_N'Y(t-1))^2}{1+\hat{r}_N'z(t)}$, and it was shown in passing in the proof of theorem 4.2 that the sequence $\{\sigma_N^2(\theta)\}$, where $\sigma_N^2(\theta) = N^{-1} \sum_{t=1}^{N} \frac{(X(t)-\beta'Y(t-1))^2}{1+r'z(t-1)}$, converges uniformly and almost surely on Θ to $\sigma^2(\theta) = E\left\{\frac{(X(t)-\beta'Y(t-1))^2}{1+r'z(t)}\right\}$. Using a similar argument to that used in the proof of theorem 4.2, it is also evident that $\hat{\sigma}_N^2$ converges almost surely to $\sigma^2(\theta_0)$, which was seen in the proof of theorem 4.1 to equal σ_0^2. Since $\hat{\gamma}_N = \hat{r}_N\hat{\sigma}_N^2$, it follows that $\hat{\gamma}_N$ converges almost surely to $r_0\sigma_0^2 = \gamma_0$. #

4.4 The Central Limit Theorem

It was seen in §4.3 that the maximum likelihood estimates required only the existence of $E(X^2(t))$ in order to be strongly consistent, unlike the least squares estimates of chapter 3 which required the existence of the fourth moments of $\{X(t)\}$. The central limit theorem of chapter 3 also required the existence of the eighth moments of $\{X(t)\}$, a condition which is not easily checked. It will be seen with respect to the maximum likelihood estimates, however, that there is a central limit theorem if the fourth moments of $\{\varepsilon(t)\}$ and $\{B(t)\}$ are finite. Again the proof will use the martingale central limit theorem of Billingsley (theorem A1.4).

In order to establish the strong consistency of the maximum likelihood estimates, the concentrated function $\ell_N(\beta,r)$ was used. As was mentioned at the end of §4.2 however, it is more convenient when proving a central limit theorem for $\hat{\beta}_N$, $\hat{\gamma}_N$ and $\hat{\sigma}^2_N$, to consider the unconcentrated function $\tilde{\ell}_N(\beta,\gamma,\sigma^2)$. Letting $\hat{\theta}_N = [\hat{\beta}_N', \hat{\gamma}_N', \hat{\sigma}^2_N]'$, $\theta_0 = [\beta_0', \gamma_0', \sigma^2_0]'$ and $\theta = [\beta', \gamma', \sigma^2]'$, we now prove the central limit theorem for $N^{\frac{1}{2}}(\hat{\theta}_N-\theta_0)$.

THEOREM 4.3. Let $\{X(t)\}$ be strictly stationary, F_t-measurable and satisfy (1.1.1) under conditions (i)-(ix). Then $N^{\frac{1}{2}}(\hat{\theta}_N-\theta_0)$ has a limiting normal distribution with mean zero and covariance matrix $I^{-1}JI^{-1}$, where I and J are derived in appendix 4.2. If $\{\varepsilon(t)\}$ and $\{B(t)\}$ are jointly normal, the covariance matrix reduces to $2I^{-1}$.

Proof. It is shown in lemma A.4.1 of appendix 4.2 that the sequence of second derivatives $\left\{\frac{\partial^2\tilde{\ell}_N(\theta)}{\partial\theta\partial\theta'}\right\}$ converges almost surely to the matrix $\frac{\partial^2\tilde{\ell}(\theta)}{\partial\theta\partial\theta'}$, where $\tilde{\ell}(\theta) = \lim_{N\to\infty}\tilde{\ell}_N(\theta)$, and that this matrix is bounded. Furthermore, it will be seen there that $\left\{\frac{\partial^2\tilde{\ell}_N(\theta)}{\partial\theta\partial\theta'}\right\}$ is uniformly convergent on a compact neighbourhood of θ_0. Now

$$\frac{\partial\tilde{\ell}_N(\hat{\theta}_N)}{\partial\theta_i} = \frac{\partial\tilde{\ell}_N(\theta_0)}{\partial\theta_i} + \left[\frac{\partial^2\tilde{\ell}_N(\tilde{\theta}_{N,i})}{\partial\theta_i\partial\theta'}\right](\hat{\theta}_N-\theta_0)$$

where θ_i is the ith component of θ and $\tilde{\theta}_{N,i}$ is on the line segment between θ_0 and $\hat{\theta}_N$. Since $\hat{\theta}_N$ converges almost surely to θ_0, then so must $\tilde{\theta}_{N,i}$ for each i. It is shown below that $N^{\frac{1}{2}}\frac{\partial\tilde{\ell}_N(\theta_0)}{\partial\theta}$ has a limiting normal distribution with mean zero and covariance matrix J, and since $\left\{\frac{\partial^2\tilde{\ell}_N(\theta)}{\partial\theta\partial\theta'}\right\}$ is uniformly convergent, $\frac{\partial^2\tilde{\ell}_N(\tilde{\theta}_{N,i})}{\partial\theta\partial\theta'}$ converges almost surely to a positive definite matrix I derived in appendix 4.2 and defined by

$$(4.4.1) \qquad I = \frac{\partial^2\tilde{\ell}(\theta_0)}{\partial\theta\partial\theta'} .$$

It is also obvious that $\frac{\partial\tilde{\ell}_N(\hat{\theta}_N)}{\partial\theta_i} = 0$, $i = 1,\ldots,n(n+3)/2$ since $\hat{\theta}_N$ converges almost surely to θ_0 which uniquely minimizes $\tilde{\ell}(\theta) = \lim_{N\to\infty}\tilde{\ell}_N(\theta)$, a function which is bounded and continuously differentiable on Θ. Thus $N^{\frac{1}{2}}(\hat{\theta}_N-\theta_0)$ will have the same asymptotic distributions as $-I^{-1}N^{\frac{1}{2}}\dfrac{\partial\tilde{\ell}_N(\theta_0)}{\partial\theta}$,

that is $N^{\frac{1}{2}}(\hat{\theta}_N-\theta_0)$ has a limiting normal distribution with mean zero and covariance matrix $I^{-1}JI^{-1}$.

Letting $u_{0t} = X(t) - \beta_0'Y(t-1)$ and $\lambda_{0t} = \sigma_0^2 + \gamma_0'z(t)$, it is seen from (4.2.4) that

$$\frac{\partial\tilde{\ell}_N(\theta_0)}{\partial\beta} = -2N^{-1}\sum_{t=1}^{N}\lambda_{0t}^{-1}u_{0t}Y(t-1),$$

$$\frac{\partial\tilde{\ell}_N(\theta_0)}{\partial\gamma} = N^{-1}\sum_{t=1}^{N}\lambda_{0t}^{-1}z(t) - N^{-1}\sum_{t=1}^{N}\lambda_{0t}^{-2}u_{0t}^2z(t)$$

and

$$\frac{\partial\tilde{\ell}_N(\theta_0)}{\partial\sigma^2} = N^{-1}\sum_{t=1}^{N}\lambda_{0t}^{-1} - N^{-1}\sum_{t=1}^{N}\lambda_{0t}^{-2}u_{0t}^2 .$$

Letting $\eta(t) = u_{0t}^2 - \lambda_{0t}$ and $\xi_t(a) = \lambda_{0t}^{-2}\{2u_{0t}\lambda_{0t}a_1'Y(t-1)+(a_3+a_2'z(t))\eta(t)\}$, where $a = [a_1',a_2',a_3]'$, a_1 and a_2 are n and $n(n+1)/2$ component vectors and a_3 is a scalar, then it is easily seen that

$$N^{-1}\sum_{t=1}^{N}\xi_t(a) = -a'\frac{\partial\tilde{\ell}_N(\theta_0)}{\partial\theta} .$$

Now

$$E(u_{0t}|F_{t-1}) = E((X(t)-\beta_0'Y(t-1))|F_{t-1}) = E((\varepsilon(t)+B(t)Y(t-1)|F_{t-1}) = 0.$$

Also, $E(\eta(t)|F_{t-1}) = E((\varepsilon(t)+B(t)Y(t-1))^2-(\sigma_0^2+\gamma_0'z(t))|F_{t-1}) = 0$, by (4.2.2). Thus $E(\xi_t(a)|F_{t-1}) = 0$. But $\xi_t(a)$ is strictly stationary and ergodic, and it will be seen shortly that $E(\xi_t^2(a))$ is finite. Hence, by theorem A.1.4, $N^{-\frac{1}{2}}\sum_{t=1}^{N}\xi_t(a)$ has a limiting normal distribution with mean zero and variance $E(\xi_t^2(a))$. This variance may be expressed in the form $a'Ja$ where J is symmetric and positive definite, and does not depend on a. Thus $N^{\frac{1}{2}}\dfrac{\partial\tilde{\ell}_N(\theta_0)}{\partial\theta}$ has a limiting normal distribution with mean zero and covariance matrix J.

It remains to be shown that $E(\xi_t^2(a))$ is finite for each a. Now

$$\xi_t^2(a) = 4u_{0t}^2\lambda_{0t}^{-2}(a_1'Y(t-1))^2+\lambda_{0t}^{-4}(a_3+a_2'z(t))^2\eta^2(t)+4\lambda_{0t}^{-3}u_{0t}a_1'Y(t-1)(a_3+a_2'z(t))\eta(t) .$$

Since $E(u_{0t}^2|F_{t-1}) = \lambda_{0t}$, the term $u_{0t}^2\lambda_{0t}^{-2}(a_1'Y(t-1))^2$ has expectation $E[(a'Y(t-1))^2\lambda_{0t}^{-1}]$ which is finite by lemma 4.2. Also, $E(\eta^2(t)|F_{t-1}) = E(u_{0t}^4|F_{t-1}) - 2\lambda_{0t}^2$, since $E(u_{0t}^2|F_{t-1}) = \lambda_{0t}$, and so $E(\eta^2(t)|F_{t-1})$ is a quartic function of Y(t-1). By lemma 4.2, therefore, $\lambda_{0t}^{-2}E(\eta^2(t)|F_{t-1})$ is bounded above and below. But $\lambda_{0t}^{-2}(a_3+a_2'z(t))^2$ is bounded above and below for the same reason, showing that $E\{\lambda_{0t}^{-4}(a_3+a_2'z(t))^2\eta^2(t)\}$ is finite, since it is equal to $E\{E[\eta^2(t)|F_{t-1}]\lambda_{0t}^{-4}(a_3+a_2'z(t))^2\}$. The third term is similarly bounded since $E(u_{0t}\eta(t)|F_{t-1}) = E\{(u_{0t}^3-u_{0t}\lambda_{0t})|F_{t-1}\} = E(u_{0t}^3|F_{t-1})$, a cubic in Y(t-1), $a_1'Y(t-1)(a_3+a_2'z(t))\lambda_{0t}^{-1}$ is bounded, and so $\lambda_{0t}^{-3}u_{0t}a_1'Y(t-1)(a_3+a_2'z(t))\eta(t)$ has finite expectation. Thus $E(\xi_t^2(a))$ is finite. #

4.5 Some Practical Aspects

In practice, having obtained the maximum likelihood estimate of (1.1.1) from (4.2.6)-(4.2.8), it will then be necessary to estimate the covariance matrix of these estimates. Such an estimate will be required for example if one wishes to carry out tests of hypotheses or derive confidence intervals for the parameters of the model.

It was seen in §4.4 that asymptotically,

$$N\ \mathrm{cov}(\hat{\theta}_N-\theta) = I^{-1}JI^{-1} ,$$

where I and J are defined in appendix 4.2. Defining the submatrices $\hat{I}_{ij}$ and $\hat{J}_{ij}$, $1 \le i \le j \le 3$ of $\hat{I}$ and $\hat{J}$ in the same way as the submatrices Ω_{ij} of Ω were defined in appendix 3.2, the obvious moment estimates $\hat{I}$ and $\hat{J}$ of I and J are given by

$$\hat{I}_{11} = 2N^{-1} \sum_{t=1}^{N} \hat{\lambda}_t^{-1} Y(t-1)Y'(t-1) \quad , \quad \hat{I}_{22} = N^{-1} \sum_{t=1}^{N} \hat{\lambda}_t^{-2} z(t)z'(t)$$

$$\hat{I}_{23} = N^{-1} \sum_{t=1}^{N} \hat{\lambda}_t^{-2} z(t) \quad , \quad \hat{I}_{33} = N^{-1} \sum_{t=1}^{N} \hat{\lambda}_t^{-2}$$

$\hat{I}_{12} = 0$ and $\hat{I}_{13} = 0$, where $\hat{\lambda} = \hat{\sigma}_N^2 + \hat{\gamma}_N' z(t)$.

Letting $\hat{u}_t = X(t) - \hat{\beta}_N' Y(t-1)$ and $\hat{\eta}_t = \hat{u}_t^2 - \hat{\lambda}_t$, then the estimates of the J_{ij} are given by

$$\hat{J}_{11} = 4N^{-1} \sum_{t=1}^{N} \hat{\lambda}_t^{-1} Y(t-1)Y'(t-1) \quad , \quad \hat{J}_{12} = 2N^{-1} \sum_{t=1}^{N} \hat{u}_t^3 \hat{\lambda}_t^{-3} Y(t-1)z'(t)$$

$$\hat{J}_{13} = 2N^{-1} \sum_{t=1}^{N} \hat{u}_t^3 \hat{\lambda}_t^{-3} Y(t-1) \quad , \quad \hat{J}_{22} = N^{-1} \sum_{t=1}^{N} \hat{\eta}_t^2 \hat{\lambda}_t^{-4} z(t)z'(t)$$

$$\hat{J}_{23} = N^{-1} \sum_{t=1}^{N} \hat{\eta}_t^2 \hat{\lambda}_t^{-4} z(t) \quad , \quad \hat{J}_{33} = N^{-1} \sum_{t=1}^{N} \hat{\eta}_t^2 \hat{\lambda}_t^{-4} .$$

It is quite a straightforward matter to show that $\hat{I}$ and $\hat{J}$ are strongly consistent for I and J.

If $\{\varepsilon(t)\}$ and $\{B(t)\}$ are jointly normally distributed then the covariance matrix of $\hat{\theta}_N$ mav be estimated by $2N^{-1}I^{-1}$, since in that case $J = 2I$, as is shown in appendix 4.2.

When carrying out tests of hypotheses on the parameters we will of course replace those parameters restricted by the null hypothesis by the values they are assumed to take under the null hypothesis. The hypothesis testing problem is considered in some detail in chapter 6.

APPENDIX 4.1

PROOF OF LEMMA 4.1. We first show that the set $A = \{\beta \in \mathbb{R}^n : \text{(ci) holds}\}$ is compact. If $\{\lambda_1,\ldots,\lambda_n\}$ are the eigenvalues of M, then it can be shown that $z^n - \sum_{j=1}^{n} \beta_j z^{n-j} = \prod_{j=1}^{n} (z-\lambda_j)$ (see Andel (1971)) and hence that β_j equals $(-1)^{j+1}$ times the sum of the products of the λ_i taken j at a time. We may thus write $\beta = f(\lambda)$, where $\beta = [\beta_1 \ldots \beta_n]'$, $\lambda = [\lambda_1 \ldots \lambda_n]'$ and f is a continuous function from $\mathbb{C}^n$ into $\mathbb{C}^n$, where $\mathbb{C}^n$ is n-dimensional complex space. If B is the compact subset of $\mathbb{C}^n$ defined by $B = \{\lambda \in \mathbb{C}^n: |\lambda_j| \leq 1-\delta_3, j = 1,\ldots,n\}$, then it is easily seen that $A = \mathbb{R}^n \cap f(B)$. However, B is compact in $\mathbb{C}^n$ and f is continuous, so that f(B) is compact in $\mathbb{C}^n$ and $\mathbb{R}^n \cap f(B)$, that is A, is compact in $\mathbb{R}^n$.

Now, for fixed $\beta \in A$, let W_β be the $n \times n$ symmetric non-negative definite matrix whose vec is the last column of $(I - M \otimes M)^{-1}$. From the proof of corollary 2.2.2, the eigenvalues of $(I - M \otimes M)^{-1}$ are $(1-\lambda_i\lambda_j)^{-1}$, $i,j = 1,\ldots,n$ where $\{\lambda_i, i = 1,\ldots,n\}$ is the set of eigenvalues of M. Thus the eigenvalues of $(I - M \otimes M)$ have moduli greater than or equal to $1 - (1-\delta_3)^2 = \delta_3(2-\delta_3)$, $|\det(I - M \otimes M)| \geq \delta_3^n(2-\delta_3)^n$ and W_β is therefore bounded element by element. Moreover, the entries of W_β are continuous in β over A, and so the smallest eigenvalue $\lambda_1(W_\beta)$ of W_β is a continuous function of β, since W_β is symmetric and non-negative definite so that its eigenvalues are real and non-negative. Since A is compact, we must therefore have $\inf_{\beta \in A} \lambda_1(W_\beta) = \lambda_1(W_{\beta^*})$, for some $\beta^* \in A$. From the proof of theorem 2.3, however, it is seen that W_β is positive definite for all $\beta \in A$ and so $\lambda_1(W_{\beta^*}) > 0$ and $\lambda_1(W_\beta)$ is bounded below by a positive number.

Let $S_\beta = \{r \in \mathbb{R}^{n(n+1)/2}: (\beta', r')' \text{ satisfies (cii) and (ciii)}$. For $r \in S_\beta$, define the $n \times n$ symmetric matrix R by $r = \text{vech } R$. Then, from Richter (1958)

$$(\text{vec } R)'(\text{vec } W_\beta) = \text{tr}(RW_\beta) \geq \sum_{i=1}^{n} \lambda_i(R)\lambda_{n-i+1}(W_\beta).$$

where $\lambda_i(D)$, $i = 1,\ldots,n$ are the eigenvalues of the symmetric matrix D ordered from minimum to maximum. Now, since the eigenvalues of R and W_β are strictly positive, we have $\lambda_n(R) \leq \delta_6/\lambda_1(W_\beta) \leq \delta_6/\lambda_1(W_{\beta*})$, and so S_β is uniformly bounded over $\beta \in A$. Thus Θ is a bounded subset of $\mathbb{R}^{n(n+3)/2}$.

To see that Θ is also closed, and therefore compact, consider the set Θ^* defined by $\Theta^* = A^* \times \mathbb{R}^{n(n+1)/2}$, where A^* is the subset of $\mathbb{R}^n$ consisting of those vectors satisfying (ci) with δ_3 replaced by δ_3' for some δ_3' such that $0 < \delta_3' < \delta_3$. Then Θ^* contains Θ and is closed since A^* is closed. If $\theta = [\beta',r']' \in \Theta^* \setminus \Theta$, where $\beta \in \mathbb{R}^n$, $r \in \mathbb{R}^{n(n+1)/2}$ and $\Theta^* \setminus \Theta$ denotes the complement of Θ in Θ^*, then $\beta \in A^* \setminus A$ or $r \in \mathbb{R}^{n(n+1)/2} \setminus S_\beta$. Thus, since $A^* \setminus A$ and $\mathbb{R}^{n(n+1)/2} \setminus S_\beta$ are both open sets, and W_β is continuous in β on A^*, there is an open neighbourhood of θ which is entirely within $\Theta^* \setminus \Theta$. Hence $\Theta^* \setminus \Theta$ is open in $\mathbb{R}^{n(n+1)/2}$, and Θ is therefore closed and compact.

#

PROOF OF LEMMA 4.2. The matrix $zz'/(1+z'\Omega z)$ is non-negative definite and is bounded above element by element if its trace is bounded. But

$$\text{tr}\{zz'/(1+z'\Omega z)\} = z'z/(1+z'\Omega z) \quad \text{and} \quad z'\Omega z \geq \lambda_1 z'z \quad .$$

Thus $0 \leq z'z/(1+z'\Omega z) \leq z'z/(1+\lambda_1 z'z) \leq \lambda_1^{-1}$ for all $z \in \mathbb{R}^p$. #

APPENDIX 4.2

LEMMA A.4.1. The sequence $\left\{\frac{\partial^2 \tilde{\ell}_N(\theta)}{\partial\theta\partial\theta'}\right\}$ converges uniformly almost surely on a compact neighbourhood of θ_0 to $\frac{\partial^2 \ell(\theta)}{\partial\theta\partial\theta'}$.

Proof. The second derivatives of $\tilde{\ell}_N(\theta)$ are given by

$$\frac{\partial^2 \tilde{\ell}_N(\theta)}{\partial\beta\partial\beta'} = 2N^{-1}\sum_{t=1}^{N} \lambda_t^{-1} Y(t-1)Y'(t-1);$$

$$\frac{\partial^2 \tilde{\ell}_N(\theta)}{\partial\beta\partial\gamma'} = 2N^{-1}\sum_{t=1}^{N} \lambda_t^{-2} u(t)Y(t-1)z'(t);$$

$$\frac{\partial^2 \tilde{\ell}_N(\theta)}{\partial\beta\partial\sigma^2} = 2N^{-1}\sum_{t=1}^{N} \lambda_t^{-2} u(t)Y(t-1);$$

$$\frac{\partial^2 \tilde{\ell}_N(\theta)}{\partial\gamma\partial\gamma'} = N^{-1}\sum_{t=1}^{N} 2\lambda_t^{-3}u^2(t)z(t)z'(t) - N^{-1}\sum_{t=1}^{N} \lambda_t^{-2} z(t)z'(t);$$

$$\frac{\partial^2 \tilde{\ell}_N(\theta)}{\partial\gamma\partial\sigma^2} = N^{-1}\sum_{t=1}^{N} 2\lambda_t^{-3}u^2(t)z(t) - N^{-1}\sum_{t=1}^{N} \lambda_t^{-2} z(t);$$

$$\frac{\partial^2 \tilde{\ell}_N(\theta)}{(\partial\sigma^2)^2} = N^{-1}\sum_{t=1}^{N} 2\lambda_t^{-3}u^2(t) - N^{-1}\sum_{t=1}^{N} \lambda_t^{-2} \quad ;$$

where $\lambda_t = \sigma^2 + \gamma' z(t)$. It follows from lemma 4.2 that the expectation of each of the above terms exists, by noting that

$$E(u(t)|F_{t-1}) = E\{(X(t)-\beta' Y(t-1))|F_{t-1}\} = (\beta_0-\beta)'Y(t-1),$$

and that

$$E(u^2(t)|F_{t-1}) = E\{(X(t)-\beta' Y(t-1))^2|F_{t-1}\} = E\{(u_{0t}+(\beta_0-\beta)'Y(t-1))^2|F_{t-1}\}$$

$$= \sigma_0^2 + \gamma_0' z(t) + [(\beta_0-\beta)' Y(t-1)]^2 ,$$

since $E(u_{0t}|F_{t-1}) = 0$. Thus $\left\{\frac{\partial^2 \tilde{\ell}_N(\theta)}{\partial\theta\partial\theta'}\right\}$ converges almost surely to a matrix which is seen to equal $\frac{\partial^2\tilde{\ell}(\theta)}{\partial\theta\partial\theta'}$, and which is given by

$$\frac{\partial^2\tilde{\ell}(\theta)}{\partial\beta\partial\beta'} = 2E[\lambda_t^{-1}Y(t-1)Y'(t-1)] ;$$

$$\frac{\partial^2\tilde{\ell}(\theta)}{\partial\beta\partial\gamma'} = 2E[\lambda_t^{-2}u(t)Y(t-1)z'(t)];$$

$$\frac{\partial^2\tilde{\ell}(\theta)}{\partial\beta\partial\sigma^2} = 2E[\lambda_t^{-2}u(t)Y(t-1)] ;$$

$$\frac{\partial^2\tilde{\ell}(\theta)}{\partial\gamma\partial\gamma'} = 2E[\lambda_t^{-3}u^2(t)z(t)z'(t)] - E[\lambda_t^{-2}z(t)z'(t)] ;$$

$$\frac{\partial^2\tilde{\ell}(\theta)}{\partial\gamma\partial\sigma^2} = 2E[\lambda_t^{-3}u^2(t)z(t)] - E[\lambda_t^{-2}z(t)];$$

$$\frac{\partial^2\tilde{\ell}(\theta)}{(\partial\sigma^2)^2} = 2E[\lambda_t^{-3}u^2(t)] - E[\lambda_t^{-2}] .$$

Furthermore, $\frac{\partial^2\tilde{\ell}(\theta)}{\partial\theta\partial\theta'}$ is obviously continuous in a compact neighbourhood $N(\theta_0)$ of θ_0, and is uniformly bounded on $N(\theta_0)$, so that $\left\{\frac{\partial^2\tilde{\ell}_N(\theta)}{\partial\theta\partial\theta'}\right\}$ converges uniformly on $N(\theta_0)$. Hence, since $\hat{\theta}_N$ converges almost surely to θ_0, $\left\{\frac{\partial^2\tilde{\ell}_N(\hat{\theta}_N)}{\partial\theta\partial\theta'}\right\}$ converges almost surely to $\frac{\partial^2\tilde{\ell}(\theta_0)}{\partial\theta\partial\theta'} = I$, whose submatrices I_{ij}, dimensionally defined analogously to the submatrices Ω_{ij} of Ω in appendix 3.2, are given by

$$I_{11} = \frac{\partial^2\tilde{\ell}(\theta_0)}{\partial\beta\partial\beta'} = 2E[\lambda_{0t}^{-1}Y(t-1)Y'(t-1)] ;$$

$$I_{12} = \frac{\partial^2\tilde{\ell}(\theta_0)}{\partial\beta\partial\gamma'} = 2E\{E[u_{0t}|F_{t-1}]\lambda_{0t}^{-2}Y(t-1)z'(t)\} = 0 ;$$

$$I_{13} = \frac{\partial^2\tilde{\ell}(\theta_0)}{\partial\beta\partial\sigma^2} = 2E[E[u_{0t}|F_{t-1}]\lambda_{0t}^{-2}Y(t-1)] = 0 ;$$

$$I_{22} = \frac{\partial^2\tilde{\ell}(\theta_0)}{\partial\gamma\partial\gamma'} = E[\lambda_{0t}^{-2}z(t)z'(t)] ;$$

$$\mathcal{I}_{23} = \frac{\partial^2\tilde{\ell}(\theta_0)}{\partial\gamma\partial\sigma^2} = E[\lambda_{0t}^{-2}z(t)]$$

$$\mathcal{I}_{33} = \frac{\partial^2\tilde{\ell}(\theta_0)}{(\partial\sigma^2)^2} = E[\lambda_{0t}^{-2}] \quad ,$$

the final three expressions being obtained by noting that $E[u_{0t}^2|F_{t-1}] = \lambda_{0t}$.

Now $\frac{\partial^2\tilde{\ell}(\theta_0)}{\partial\beta\partial\beta'}$ is obviously positive definite, for otherwise there would exist an n-component vector a with a'Y(t-1) = 0 almost surely in violation of assumption (iv) (see §2.1). Also

$$\begin{bmatrix} \frac{\partial^2\tilde{\ell}(\theta_0)}{\partial\gamma\partial\gamma'} & \frac{\partial^2\tilde{\ell}(\theta_0)}{\partial\gamma\partial\sigma^2} \\ \frac{\partial^2\tilde{\ell}(\theta_0)}{\partial\sigma^2\partial\gamma'} & \frac{\partial^2\tilde{\ell}(\theta_0)}{(\partial\sigma^2)^2} \end{bmatrix} = E\left\{\lambda_{0t}^{-2}\begin{bmatrix} z(t) \\ 1 \end{bmatrix}[z'(t)\ 1]\right\}$$

which must also be positive definite, since by lemma 3.1 there are no constant c and n(n+1)/2-component vector a such that a'z(t) + c = 0 almost surely. Hence $\mathcal{I}$ is positive definite.

The Derivation of J

The matrix J defined in theorem 4.3 is obtained by expressing $E(\xi_t^2(a))$ in the form a'Ja, where J does not depend on a, and is symmetric. Letting

$$J = \begin{bmatrix} J_{11} & J_{12} & J_{13} \\ J_{12}' & J_{22} & J_{23} \\ J_{13}' & J_{23}' & J_{33} \end{bmatrix}$$

where J_{ij} is an n(i) × n(j) matrix, n(1) = n, n(2) = n(n+1)/2 and n(3) = 1, J_{ij} may be found by evaluating the component of $E(\xi_t^2(a))$ of the form $a_i'J_{ij}a_j$. Thus

$$J_{11} = 4E\{E[u_{0t}^2|F_{t-1}]\lambda_{0t}^{-2}Y(t-1)Y'(t-1)\} = 4E\{Y(t-1)Y'(t-1)\lambda_{0t}^{-1}\} ;$$

$$J_{12} = 2E\{E[u_{0t}\eta(t)|F_{t-1}]\lambda_{0t}^{-3}Y(t-1)z'(t)\} = 2E\{u_{0t}^3\lambda_{0t}^{-3}Y(t-1)z'(t)\} ;$$

$$J_{13} = 2E\{E[u_{0t}\eta(t)|F_{t-1}]\lambda_{0t}^{-3}Y(t-1)\} = 2E\{u_{0t}^3\lambda_{0t}^{-3}Y(t-1)\} ;$$

$$J_{22} = E\{\eta^2(t)\lambda_{0t}^{-4}z(t)z'(t)\}; \quad J_{23} = E\{\eta^2(t)\lambda_{0t}^{-4}z(t)\} ; \quad J_{33} = E\{\eta^2(t)\lambda_{0t}^{-4}\} .$$

If $\{\varepsilon(t)\}$ and $\{B(t)\}$ are jointly normal, then u_{0t}, conditional on F_{t-1}, is distributed normally with mean zero and variance $(\sigma_0^2+\gamma_0'z(t))$. Since $\eta(t) = u_{0t}^2 - (\sigma_0^2+\gamma_0'z(t))$, it therefore follows that $E(u_{0t}\eta(t)|F_{t-1}) = E\{[u_{0t}^3-u_{ot}(\sigma_0^2+\gamma_0'z(t))]|F_{t-1}\} = 0$. Furthermore, in this case $E(u_{0t}^4|F_{t-1}) = 3\lambda_{0t}^2$, and so $E(\eta^2(t)|F_{t-1}) = 2(\sigma_0^2+\gamma_0'z(t))^2 = 2\lambda_{0t}^2$. Thus the matrix J may be simplified to give $J_{12} = 0$, $J_{13} = 0$, $J_{22} = E\{2\lambda_{0t}^{-2}z(t)z'(t)\}$, $J_{23} = E\{2\lambda_{0t}^{-2}z(t)\}$ and $J_{33} = E\{2\lambda_{0t}^{-2}\}$, and $J = 2I$, giving $I^{-1}JI^{-1} = 2I^{-1}$.

Thus, if $\{\varepsilon(t)\}$ and $\{B(t)\}$ are normal, the asymptotic covariance matrices of $N^{\frac{1}{2}}(\hat{\beta}_N-\beta_0)$, $N^{\frac{1}{2}}(\hat{\gamma}_N-\gamma_0)$ and $N^{\frac{1}{2}}(\hat{\sigma}_N^2-\sigma_0^2)$ are determined in a straightforward manner using the formula for the partitioned inverse of a matrix. The asymptotic covariance matrix of $N^{\frac{1}{2}}(\hat{\beta}_N-\beta_0)$ is $[E\{\lambda_{0t}^{-1}Y(t-1)Y'(t-1)\}]^{-1}$, while that of $N^{\frac{1}{2}}(\hat{\gamma}_N-\gamma_0)$ is given by

$$2\{E[\lambda_{0t}^{-2}z(t)z'(t)]-E[\lambda_{0t}^{-2}z(t)](E[\lambda_{0t}^{-2}])^{-1}E[\lambda_{0t}^{-2}z'(t)]\}^{-1}$$

and the asymptotic variance of $N^{\frac{1}{2}}(\hat{\sigma}_N^2-\sigma_0^2)$ is

$$2\{E[\lambda_{0t}^{-2}]-E[\lambda_{0t}^{-2}z'(t)](E[\lambda_{0t}^{-2}z(t)z'(t)])^{-1}E[\lambda_{0t}^{-2}z(t)]\}^{-1}$$

CHAPTER 5

A MONTE CARLO STUDY

5.1 Simulation and Estimation Procedures

In order to illustrate the procedures introduced in chapters 3 and 4 a number of simulations were performed with first and second order univariate RCA models for several sets of data of different sizes. While the simulations performed have been by no means exhaustive, as we shall see the results do conform with the asymptotic theory developed in the last two chapters.

When generating series to be used in the application of the methods of estimation, condition (vi) of chapter 3, which is given in an alternative form in chapter 4, must be satisfied for the estimation procedure to be sure of working. In the light of this we use the following method to generate random coefficient autoregressions for which both $\{\varepsilon(t)\}$ and $\{B(t)\}$ are sequences of normally distributed random variables.

1. Specify the real and complex eigenvalues $\{\lambda_i,\ i = 1,\ldots,n\}$ of the matrix M, which must all have moduli less than unity.

2. Calculate the parameters $\{\beta_i;\ i = 1,\ldots,n\}$ from

$$\prod_{i=1}^{n} (1 - \lambda_i z) = 1 - \sum_{i=1}^{n} \beta_i z^i \quad .$$

Thence calculate the square matrix W where $\operatorname{vec} W$ is the last column of $(I - M \otimes M)^{-1}$.

3. Take a positive definite matrix Σ^* and compute $\operatorname{tr}(\Sigma^* W)$.

Now in order to guarantee second order stationarity we must have

$$1 > (\operatorname{vec} \Sigma)' \operatorname{vec} W = \operatorname{tr}(\Sigma W),$$

so that if we specify the value of $\rho = \operatorname{tr}(\Sigma W)$ required, then

$\Sigma = \Sigma^*(\rho/tr(\Sigma^* W))$ satisfies $tr(\Sigma W) = \rho$. Thus we need only define Σ^* up to a multiplicative constant.

4. Compute the lower triangular matrix L which has positive diagonal elements and for which $LL' = \Sigma$.

5. Generate a vector $[\omega(1), \omega(2),\ldots,\omega(n + 1)]'$ where the $\omega(i)$'s are successive generates of a standard normal random number generator. Specifying $\sigma^2 = E\{\varepsilon^2(t)\}$, let $\varepsilon(t) = \sigma\ \omega(1)$ and $B'(t) = L[\omega(2),\ldots,\omega(n+1)]'$. Then $\varepsilon(t)$ and $B(t)$ will theoretically be independent, have zero means and have $E\{\varepsilon^2(t)\} = \sigma^2$ and $E\{B'(t)B(t)\} = LL' = \Sigma$.

6. Calculate

$$X(t) = \sum_{i=1}^{n} (\beta_i + B_i(t))X(t-i) + \varepsilon(t)\ , \qquad \text{where} \quad X(t) = 0$$

for $t < 0$,

7. Repeat steps 5 and 6 $(N + 200)$ times where N is the sample size desired, and ignore the first 200 values produced. This enables $\{X(t)\}$ to reach an equilibrium since, under (vi), $\{X(t)\}$ is stable.

A realization of two hundred observations for each of four first order random coefficient autoregressive models was generated, i.e. models of the form

$$(5.1.1) \qquad X(t) = (\beta + B(t))\ X(t - 1) + \varepsilon(t),$$

where the $\varepsilon(t)$ are normally and independently distributed random variables each with mean zero and variance σ^2, i.e. $N.I.D.(0,\sigma^2)$, while the $B(t)$ are $N.I.D.(0,\delta^2)$. For this model $\gamma = \delta^2$. In each case the concentrated log likelihood function $\ell_N(\beta,r)$, where $r = \gamma/\sigma^2$, over the subset $\mathbb{R}^2$ defined by $|\beta| \le 2$ and $0 \le r \le 4$ was computed. The three dimensional graphs of $-\ell_N(\beta,r)$ are depicted in figures 5.1 - 5.4, with two perspectives of each surface being given in each case.

FIGURE 5.1 Log-likelihood $-\ell_N(\beta, r)$

$\beta = .8$, $\delta^2 = .25$, $\sigma^2 = 1.0$

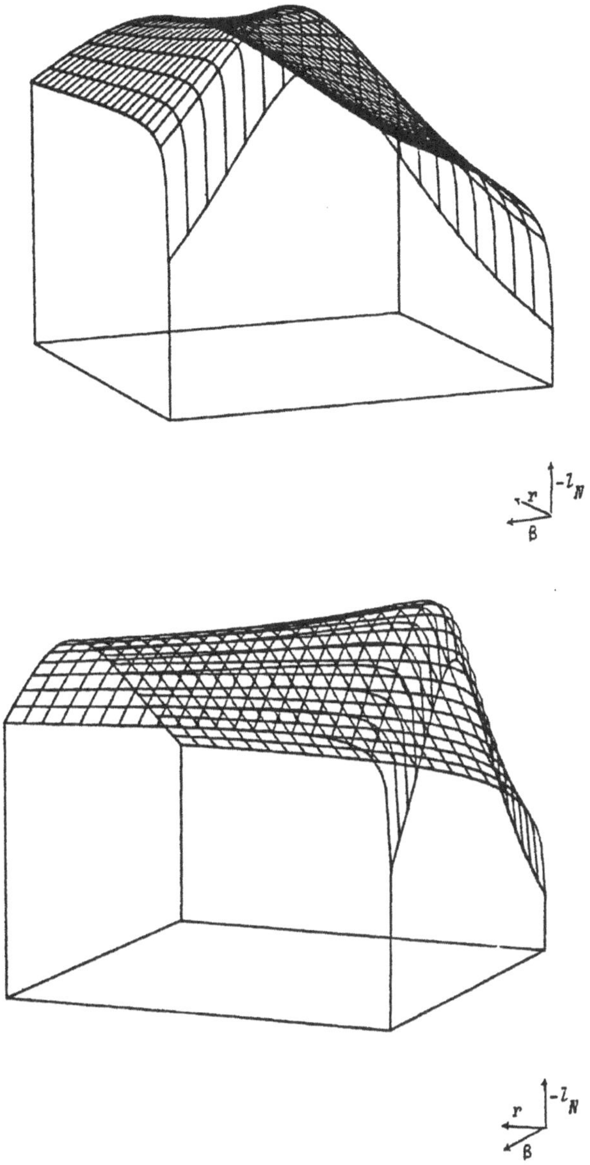

FIGURE 5.2 Log-likelihood $-\ell_N(\beta, r)$

$\beta = .0$, $\delta^2 = .81$, $\sigma^2 = 1.0$

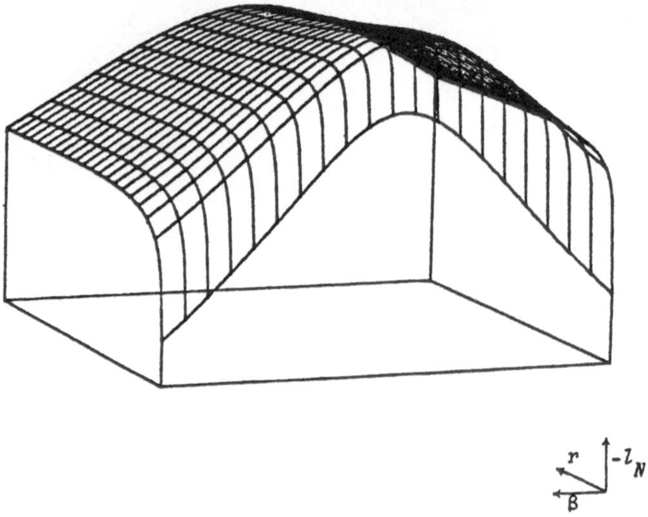

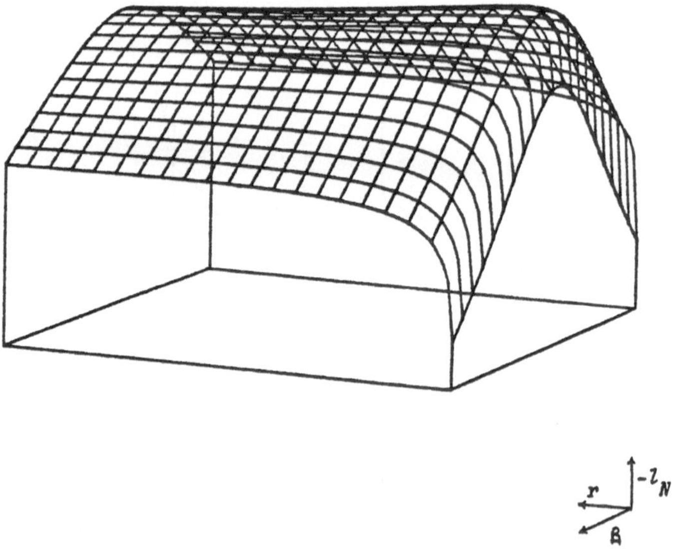

FIGURE 5.3 Log-likelihood $-\ell_N(\beta, r)$

$\beta = .5$, $\delta^2 = .25$, $\sigma^2 = 1.0$

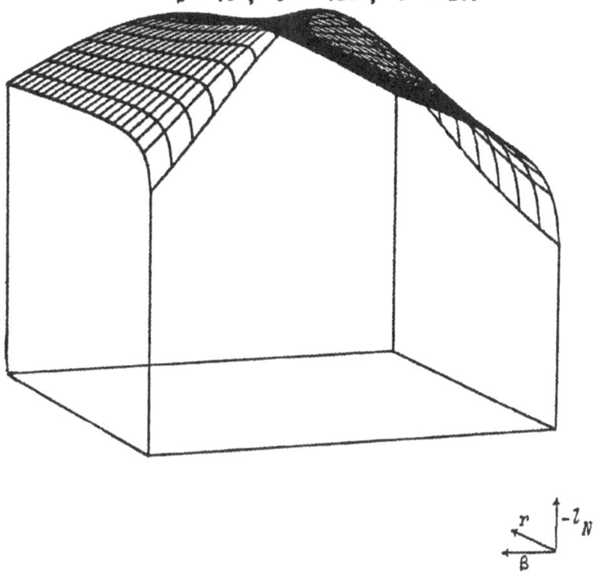

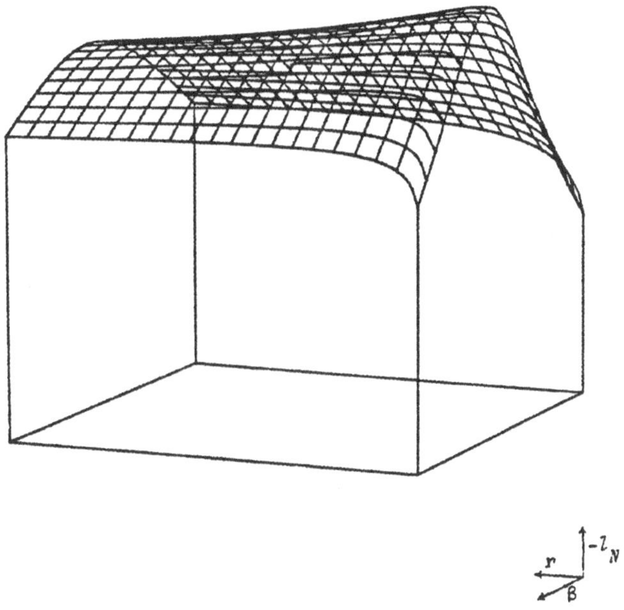

FIGURE 5.4 Log-likelihood $-\ell_N(\beta, r)$

$\beta = -.9$, $\delta^2 = .16$, $\sigma^2 = 1.0$

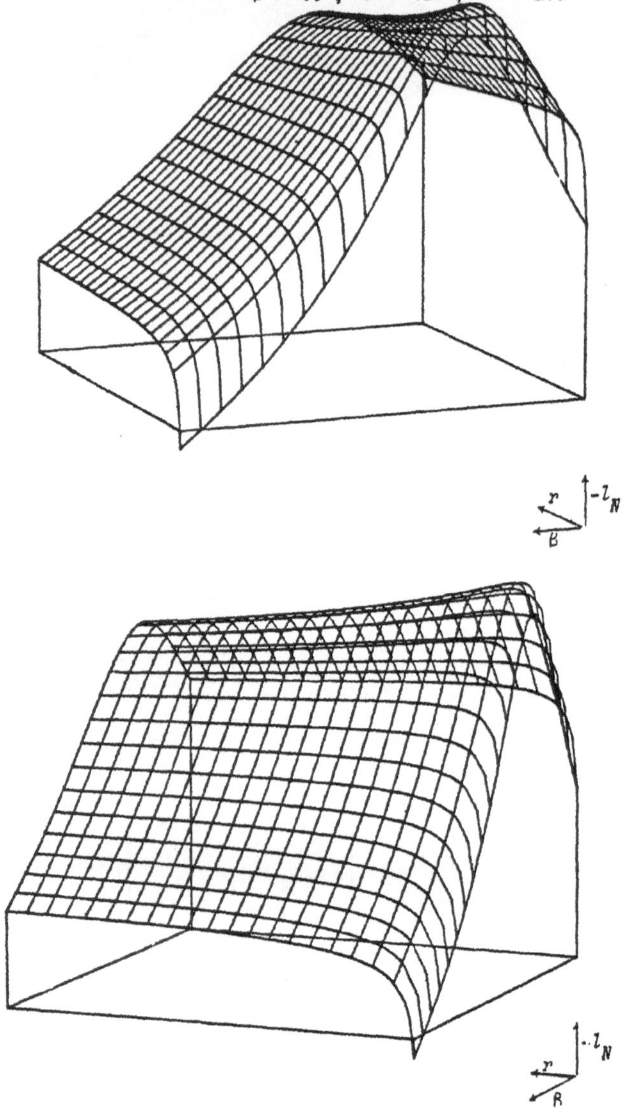

From these graphs we see that although the maximum likelihood estimate $\hat{\beta}_N$ of β_0 is well pronounced, the likelihood will be relatively flat in the r direction, so that some care must be taken in assuring that an iterative maximum likelihood procedure has actually converged.

The method we use to obtain the maximum likelihood estimates is the variable step-length Newton-Raphson method. This method requires an initial estimate $\theta^{(0)}$ of the estimates to commence the iterative procedure. There is a distinct advantage in being able to start the iterative procedure with a strongly consistent estimate of the parameters since then we are commencing the iterations close to the global optimum. This is of importance when we have likelihoods such as those appearing in figures 5.1 - 5.4 where the likelihood is relatively flat in one direction so that convergence is slow. In such cases, starting the iterations a long way from the global optimum will result in a large number of iterations being required. One other case which might occur is that where there are a number of local optima in the likelihood surface and, by commencing the iterations far from the global optimum, we may get convergence to a local rather than the global optimum or divergence. To minimize the possibility of such situations arising we commence the Newton-Raphson algorithm with a strongly consistent estimate $\theta^{(0)}$ obtained by the least squares procedure of Chapter 3.

The maximum likelihood estimate $\hat{\theta}_N$ of θ_0 is obtained by the following procedure. Letting $\theta^{(j)}$ be the jth iterate produced by the method, then $\theta^{(j+1)}$ is obtained from

$$\theta_{j+1}(\lambda) = \theta^{(j)} - \lambda \left\{\frac{\partial^2 \ell_N(\theta^{(j)})}{\partial\theta\partial\theta'}\right\}^{-1} \frac{\partial \ell_N(\theta^{(j)})}{\partial\theta}, \ \lambda = 2^{-k}, \ k = 0,\ldots, 8,$$

$$k^* = \min \{k : 0 \leq k \leq 8, \ \ell_N(\theta_{j+1}(2^{-k})) < \ell_N(\theta^{(j)})\}$$

$$\theta^{(j+1)} = \theta_{j+1}(k^*).$$

Iteration is halted when each of the elements of $(\theta^{(j+1)} - \theta^{(j)})$ has absolute value sufficiently small, the value used throughout our experiments being 10^{-6}.

The usual Newton-Raphson method considers only the case where $\lambda = 1$. However, there is a possibility that the initial estimate and a number of early iterations may be so far from $\hat{\theta}_N$ that further iterations "overshoot" $\hat{\theta}_N$ and either converge to some other point or diverge. By allowing λ to decrease and by using the above technique, we are ensuring that the function $\ell_N(\theta^{(j)})$ is decreasing with j, and thus we can be more certain that the iterations $\theta^{(j)}$ are converging to $\hat{\theta}_N$. Since $\mathbb{R}^{n(n+1)/2}$ is a complete metric space, that is, each Cauchy sequence converges, the sequence $\{\theta^{(j)}\}$ may be taken as having converged to a minimum of the function $\ell_N(\theta)$, which would be the absolute minimum if the initial estimate $\theta^{(0)}$ were close enough to $\hat{\theta}_N$.

5.2 First and Second Order Random Coefficient Autoregressions

We consider in this section the simulation and estimation of four random coefficient autoregressions - two first order and two second order models. In light of the fourth moment condition on $\{X(t)\}$ which is required for the strong consistency of the least squares estimates, we now obtain a condition on the parameters β and δ^2 of (5.1.1) which ensures the finiteness of $E(X^4(t))$. This condition will be used to generate two first order series, only one of which satisfies the condition.

THEOREM 5.1 An F_t-measurable stationary solution to (5.1.1), with $\{\varepsilon(t)\}$ and $\{B(t)\}$ normal, exists and has finite fourth moments if and only if $\beta^4 + 6\beta^2\delta^2 + 3\delta^4 < 1$.

Proof. It was shown at the end of §2.3 that a solution exists if and only if $\beta^2 + \delta^2 < 1$, and from (2.3.2) is seen to be given by

$$X(t) = \varepsilon(t) + \sum_{j=1}^{\infty} D(t,j)\varepsilon(t-j) \tag{5.2.1}$$

where $D(t,j) = \prod_{k=0}^{j-1} \{\beta + B(t-1)\}$. Thus, if $E(X^4(t)) < \infty$, then from (5.2)

$$E(X^4(t)) = E(\varepsilon^4(t)) + 6E(\varepsilon^2(t)) \sum_{j=1}^{\infty} E(D^2(t,j))E(\varepsilon^2(t-j)) + E\{\sum_{j=1}^{\infty} D(t,j)\varepsilon(t-j)\}^4 \quad (5.3.1)$$

The first two terms are finite since $\{\varepsilon(t)\}$ is normal and $\{X(t)\}$ has finite second moments. However

$$E\{\sum_{j=1}^{\infty} D(t,j)\varepsilon(t-j)\}^4$$

$$= E\{\sum_{j=1}^{\infty} D^4(t,j)\varepsilon^4(t-j)\} + 6E\{\sum_{j<k}^{\infty} D^2(t,j)D^2(t,k)\varepsilon^2(t-j)\varepsilon^2(t-k)\}$$

$$\geq E[\varepsilon^4(t)] \sum_{j=1}^{\infty} [E\{\beta+B(t)\}^4]^j$$

$$= E[\varepsilon^4(t)] \sum_{j=1}^{\infty} [\beta^4+6\beta^2\delta^2+3\delta^4]^j$$

showing that the condition that $(\beta^4+6\beta^2\delta^2+3\delta^4)$ be less than 1 is necessary. Moreover, for $j < k$,

$$E\{D^2(t,j)D^2(t,k)\} = E\{D^4(t,j)\}\, E[\prod_{\ell=j+1}^{k} \{\beta+B(t-\ell)\}^2] .$$

Thus

$$E\{\sum_{j<k}^{\infty} D^2(t,j)D^2(t,k)\varepsilon^2(t-j)\varepsilon^2(t-k)\}$$

$$= \{E[\varepsilon^2(t)]\}^2 \sum_{j=1}^{\infty} [\beta^4+6\beta^2\delta^2+3\delta^4]^j \sum_{k=j+1}^{\infty} (\beta^2+\delta^2)^{k-j}$$

which, since $\sum_{k=j+1}^{\infty} (\beta^2+\delta^2)^{k-j} = (\beta^2+\delta^2)/(1-\beta^2-\delta^2)$, will also converge if

$(\beta^4+6\beta^2\delta^2+3\delta^4) < 1$. Hence, noting that $(\beta^2+\delta^2)^2 < \beta^4+6\beta^2\delta^2+3\delta^4$,

the condition is seen to be both necessary and sufficient. #

If $\beta = .5$ and $\delta^2 = .25$, the above theorem shows that $E(X^4(t)) < \infty$, while if $\beta = -.9$ and $\delta^2 = .16$, the fourth moments of $\{X(t)\}$ will not exist. These are the two first order cases which we consider in order to highlight the fact that it was found necessary to assume the finiteness of the fourth moments of $\{X(t)\}$ to demonstrate the strong consistency of the least squares estimation procedure. For each case the process was simulated with $\sigma^2 = E(\varepsilon^2(t)) = 1$ for the sample sizes 50, 100 and 500, for each of which one hundred replications were performed. Least squares and maximum likelihood estimates of β, δ^2 and σ^2 were then computed for each sample, and various statistics were calculated over the sets of one hundred replications. These results are summarized in table 5.1, in which lines (a) contain the true parameters, lines (b) the averages of the various estimates, lines (c) the sample variances and lines (d) the sample mean square errors. Thus the estimated standard deviations of the averages of the estimates may be obtained by taking the square roots of the values in lines (c) and dividing these by $\sqrt{100} = 10$.

TABLE 5.1

SIMULATION RESULTS FOR TWO FIRST ORDER RCA's

Sample Size		β	δ^2	σ^2	β	δ^2	σ^2
		Least Squares			Maximum Likelihood		
I	(a)	.5	.25	1.0	.5	.25	1.0
50	(b)	.4693	.1323	1.1614	.4845	.1872	1.0750
	(c)	.0220	.0233	.0940	.0230	.0336	.0998
	(d)	.0229	.0371	.1201	.0233	.0375	.1054
100	(b)	.5066	.1558	1.1196	.5241	.2080	1.0250
	(c)	.0105	.0131	.0728	.0094	.0172	.0506
	(d)	.0105	.0220	.0871	.0100	.0190	.0513
500	(b)	.4988	.2061	1.0828	.4999	.2511	1.0014
	(c)	.0031	.0089	.0289	.0021	.0034	.0070
	(d)	.0031	.0108	.0357	.0022	.0034	.0070
II	(a)	-.9	.16	1.0	-.9	.16	1.0
50	(b)	-.8255	.1109	1.2069	-.8757	.1407	1.0554
	(c)	.0082	.0100	.2084	.0086	.0078	.0906
	(d)	.0138	.0124	.2512	.0092	.0082	.0937
100	(b)	-.8247	.1305	1.1679	-.8760	.1463	1.0547
	(c)	.0038	.0071	.1770	.0040	.0042	.0550
	(d)	.0095	.0080	.2052	.0045	.0043	.0580
500	(b)	-.8581	.1324	1.1262	-.8874	.1532	1.0022
	(c)	.0009	.0030	.1346	.0009	.0009	.0082
	(d)	.0026	.0037	.1505	.0011	.0010	.0082

As is to be expected, the maximum likelihood estimation procedure has performed better, as is seen from the fact that the sample mean square errors are generally smaller than for the least squares procedure. The results for the least squares estimates are also seen to agree with the asymptotic results of chapter 3, since for the first model the sample variances decrease (the estimates are strongly consistent) while for the second model, even with a sample of size 500, the estimates of δ^2 and σ^2

are evidently by no means reliable. Moreover, it is simple to show using the techniques of the proof of theorem 5.1 that the solutions of (5.1.1) will not have finite eighth moments unless $E((\beta+B(t))^8) < 1$. It is not hard to see that neither of the models which we have considered generates processes with finite eighth moments, so that a central limit theorem for the least squares estimates of δ^2 and σ^2 cannot be expected to exist. This is borne out by the sizes of the standard deviations. For example, for the first model with sample size 500, the estimated standard deviation of the average of the least squares estimates of σ^2 is $(.0289)^{\frac{1}{2}}/10 = .017$, so that the average deviates from the true value by $82.8/17 \simeq 4.9$ standard deviations.

From the theory developed in chapter 4 we see that these higher order moment conditions were not required to obtain the asymptotic results for the maximum likelihood estimates. Indeed, for the case just discussed, the average of the maximum likelihood estimates of σ^2 deviates from the true value by $.0014{\times}10/(.0070)^{\frac{1}{2}} \simeq .17$ standard deviations. The only results which are not consistent with the theory are those for the maximum likelihood estimates of δ^2 for the second model with sample sizes 100 and 500. In each of these cases, the average of the maximum likelihood estimates of δ^2 differs from the true value by slightly more than two standard deviations, which should give no real course for alarm.

It is more difficult to obtain conditions which ensure the existence of the fourth moments of scalar processes of order larger than one. We have nevertheless used "one-off" techniques to choose two scalar second order models of the form (1.1.1), the solution of the first of which (model I), with $\beta_1 = 0$, $\beta_2 = .36$, $\Sigma_{11} = \Sigma_{22} = .2176$, $\Sigma_{12} = 0$ and $\sigma^2 = 1$, has finite fourth moments, while the solution of the second (model II) does not have these moments finite. For this model $\beta_1 = .8$, $\beta_2 = -.15$, $\Sigma_{11} = \Sigma_{12} = .0919$, $\Sigma_{22} = .1838$ and $\sigma^2 = 1$. It should be remembered that $\Sigma = E[B'(t)B(t)]$, so that, for example $\Sigma_{22} = E(B_1^2(t))$. The first of these models has $tr(\Sigma W) = .5$, while the eigenvalues of M are both .6. The second has $tr(\Sigma W) = .8$, while the eigenvalues of M are .3 and .5. In light of condition (vi) assumed in chapters 3 and 4, therefore, it is not hard to see why the fourth moments of the solution of the second model might not be finite. As was the case with the first order models, neither solution has finite eighth moments, so that, although the strong consistency of the least squares estimates of Σ and σ^2 is ensured, a central limit theorem for these estimates cannot be expected to hold. Each model was used to generate one hundred replications each of samples of length 100, 200, 500 and 1000. Various statistics were computed and are recorded in tables 5.2 and 5.3, where lines (a) contain the true parameter values, while lines (b), (c) and (d) contain the sample means, sample variances and mean-square errors respectively over the one hundred replications of the least squares and maximum likelihood estimation procedures in each case.

For each sample length the results are not unexpected. The averages of the least squares estimates of each of the parameters for the first model are converging with increasing sample size to the true values although for the parameters Σ and σ^2 they are several standard errors from these values. Since the fourth moments of the solution of the first model are finite, a central limit theorem exists for the estimates of β_1 and β_2, for which there is practical agreement indicated by the results. For example, for sample size 1000, the average of the least-squares estimates of β_2 is .3578 while the estimated standard deviation is $(.0015)^{\frac{1}{2}}/10 \simeq .004$ and the true value of β_2 is .36. The maximum likelihood estimates are seen to perform better than the least squares estimates, as indicated by the respective mean square errors.

Since the solution of the second model does <u>not</u> have finite fourth moments, the least squares estimates of Σ and σ^2 should not behave very well, which is borne out by an examination of the averages of these estimates recorded in lines (b) in table 5.2. The averages of the estimates of β_1 and β_2, however, appear to be converging to their true values at a rate which is unexpected since it was necessary to assume finite fourth moments to obtain a central limit theorem for these estimates. The only results for the maximum likelihood estimates which require some comment are those for the estimates of Σ_{22} in the case of model II. For each sample size the average of the estimates has overestimated the true value .1838 by between two and three estimated standard deviations. A possible explanation for this is that for some of the replications, the likelihood may have been particularly flat in the r-direction, and consequently our procedure may not have converged to the global maximum of the likelihood function.

TABLE 5.2

SIMULATION RESULTS FOR MODEL I

Sample Size		β_1	β_2	Σ_{11}	Σ_{12}	Σ_{22}	σ^2
	(a)	.0	.36	.2176	.0	.2176	1.0
				Least Squares			
100	(b)	.0	.3191	.1688	.0045	.1835	1.2369
	(c)	.0185	.0145	.0187	.0102	.0186	.2689
	(d)	.0185	.0162	.0211	.0102	.0198	.3250
				Maximum Likelihood			
	(b)	.0068	.3486	.2211	.0069	.2377	.9824
	(c)	.0147	.0130	.0150	.0089	.0157	.0628
	(d)	.0147	.0131	.0150	.0089	.0161	.0631
				Least Squares			
200	(b)	.0042	.3389	.1549	-.0010	.1593	1.2390
	(c)	.0061	.0072	.0102	.0058	.0095	.0777
	(d)	.0061	.0076	.0141	.0058	.0129	.1348
				Maximum Likelihood			
	(b)	.0047	.3493	.2068	.0050	.2106	1.0287
	(c)	.0050	.0057	.0080	.0055	.0108	.0334
	(d)	.0050	.0058	.0081	.0055	.0108	.0342
				Least Squares			
500	(b)	-.0035	.3492	.1828	.0063	.1848	1.1454
	(c)	.0040	.0031	.0073	.0055	.0099	.0493
	(d)	.0040	.0032	.0085	.0055	.0110	.0704
				Maximum Likelihood			
	(b)	.0012	.3558	.2083	.0022	.2192	1.0134
	(c)	.0026	.0023	.0038	.0019	.0042	.0144
	(d)	.0026	.0023	.0039	.0019	.0042	.0146
				Least Squares			
1000	(b)	-.0039	.3578	.1874	-.0009	.1975	1.1111
	(c)	.0020	.0015	.0050	.0048	.0067	.0318
	(d)	.0020	.0015	.0059	.0048	.0071	.0441
				Maximum Likelihood			
	(b)	.0013	.3607	.2185	-.0003	.2244	.9876
	(c)	.0012	.0010	.0018	.0009	.0019	.0053
	(d)	.0012	.0010	.0018	.0009	.0019	.0055

TABLE 5.3

SIMULATION RESULTS FOR MODEL II

Sample Size		β_1	β_2	Σ_{11}	Σ_{12}	Σ_{22}	σ^2
	(a)	.80	-.15	.0919	.0919	.1838	1.0
				Least Squares			
100	(b)	.7562	.1585	.0836	.0377	.1325	1.7901
	(c)	.0204	.0189	.0246	.0410	.0563	2.7051
	(d)	.0223	.0190	.0247	.0439	.0589	3.3294
				Maximum Likelihood			
	(b)	.8195	-.1663	.1386	.0294	.2276	.9101
	(c)	.0140	.0093	.0089	.0080	.0125	.0539
	(d)	.0144	.0096	.0111	.0119	.0144	.0620
				Least Squares			
200	(b)	.7617	-.1774	.0828	.0659	.1207	1.6560
	(c)	.0107	.0088	.0124	.0118	.0322	1.8306
	(d)	.0122	.0096	.0125	.0125	.0362	2.2610
				Maximum Likelihood			
	(b)	.7869	-.1556	.1173	.0546	.2137	.9516
	(c)	.0048	.0039	.0061	.0046	.0079	.0251
	(d)	.0050	.0039	.0067	.0060	.0088	.0274
				Least Squares			
500	(b)	.7785	-.1615	.0804	.0716	.1066	1.7317
	(c)	.0066	.0061	.0100	.0121	.0282	.9164
	(d)	.0071	.0062	.0101	.0125	.0342	1.4518
				Maximum Likelihood			
	(b)	.7957	-.1471	.0972	.0830	.1992	.9782
	(c)	.0019	.0017	.0013	.0008	.0026	.0127
	(d)	.0019	.0017	.0013	.0009	.0028	.0132
				Least Squares			
1000	(b)	.7815	-.1613	.0885	.0686	.1358	1.5933
	(c)	.0055	.0042	.0092	.0107	.0307	1.2692
	(d)	.0058	.0045	.0092	.0112	.0330	1.6212
				Maximum Likelihood			
	(b)	.8016	-.1497	.0949	.0909	.1921	.9873
	(c)	.0010	.0008	.0006	.0004	.0013	.0060
	(d)	.0010	.0008	.0006	.0004	.0014	.0062

5.3 Summary

The numerical results presented in this chapter are for first and second order scalar models and illustrate the theoretical results obtained in chapters 3 and 4. One fact that does emerge is that as the order of models increases the confirming of the moment conditions required for strong consistency and the validity of the central limit theorems is tedious. This is not restricted to RCA models however, as other classes of non-linear models would experience similar difficulties.

For the scalar models we have considered here, as figures 5.1 - 5.4 show, the likelihood surface is rather flat in one direction. From the point of view of computational efficiency the choice of an algorithm which will give convergence to the optimum in a reasonable period of time is important. Indeed in the case of the second order processes, the results of which were reported in the previous section, the simulation and estimation of such processes involved around 35 minutes of C.P.U. time on a Univac U1100 computer. With modern technology moving at such a fast rate however this may not be an important consideration in the future.

CHAPTER 6

TESTING THE RANDOMNESS OF THE COEFFICIENTS

6.1 Introduction

As indicated in the previous chapter, the asymptotic results for the maximum likelihood estimates may be used to test certain hypotheses of interest. The condition (cii), however, which was assumed so as to obtain a standard central limit theorem for the maximum likelihood estimates, precludes the use of the theory derived in chapter 4 to test what is perhaps the most relevant hypothesis, namely that $\Sigma = 0$, that is, that the data come from a fixed coefficient autoregression. This chapter examines the testing of hypotheses in general, and in particular, two tests for the hypothesis that $\Sigma = 0$.

The tests which we propose are based on a consideration of the likelihood ratio criterion and a statistic involving scores. Under the conditions of chapter 4, it will be shown that the two statistics are equivalent, and may be used in the testing of hypotheses, even though the likelihood is not the true likelihood in that we have used the Gaussian likelihood without having assumed normality. Moreover, the score statistic may be used in testing the hypothesis that $\Sigma = 0$, unlike the likelihood ratio statistic, whose distributional properties are difficult to obtain when $\Sigma = 0$. This is not surprising since technical difficulties usually arise when the true parameter lies on the natural boundary of the parameter space. Moran (1971) and Chant (1974) have discussed estimation and testing in such non-standard situations. Moran notes that while there is generally a problem associated with finding the distribution of the maximum likelihood estimates, there is usually no such problem with respect to the analysis of the $C(\alpha)$ tests of Neyman (1959), which involve statistics computed from vectors of scores.

The first test which we shall consider is the $C(\alpha)$ test suitably modified to account for the fact that the likelihood function may not be the true one, while the second test recovers some of the power lost by the first.

6.2 The Score Test

Before developing a test for the hypothesis that $\Sigma = 0$ (or equivalently that $\gamma = 0$) we shall consider a test for $\gamma = \gamma_o$ where the vector $[\beta_o', r_o']'$ lies interior to Θ. The likelihood ratio statistic τ_N is given by

$$(6.2.1) \qquad \tau_N = N\{\tilde{\ell}_N(\tilde{\beta}_N,\gamma_o,\tilde{\sigma}_N^2) - \tilde{\ell}_N(\hat{\beta}_N,\hat{\gamma}_N,\hat{\sigma}_N^2)\}$$

where

$$\tilde{\ell}_N(\tilde{\beta}_N,\gamma_o,\tilde{\sigma}_N^2) = \inf_{\beta,\sigma^2} \tilde{\ell}_N(\beta,\gamma_o,\sigma^2) .$$

In determining the asymptotic distribution of τ_N it is usual to derive an equivalent expression in terms of the derivatives of $\tilde{\ell}_N$. Letting $\theta_o = (\beta_o', \gamma_o', \sigma_o^2)'$, $\theta = (\beta',\gamma',\sigma^2)'$ and $I = \lim_{N\to\infty} \dfrac{\partial^2\tilde{\ell}_N(\theta_o)}{\partial\theta\partial\theta'}$, in the notation of §4.4, we have the following result.

LEMMA 6.1 <u>Let</u>

$$\tilde{\tau}_N = N/2 \, \frac{\partial\tilde{\ell}_N(\theta_o)}{\partial\theta'} \, (I^{-1}-\tilde{I}) \, \frac{\partial\tilde{\ell}_N(\theta_o)}{\partial\theta},$$

<u>where the block diagonal matrix</u> $\tilde{I}$ = <u>diag</u>$\{I_{11}^{-1} \vdots 0 \vdots I_{33}^{-1}\}$. <u>Then</u> $\tau_N - \tilde{\tau}_N \overset{p}{\to} 0$.

<u>Proof.</u> See appendix 6.1.

We shall now examine the matrix $I^{-1}-\tilde{I}$ and consider the implications of the fact that $\tilde{\ell}_N$ is not obtained from the true likelihood function. From appendix 4.2 we have

$$I^{-1} = \begin{bmatrix} I_{11}^{-1} & 0 & 0 \\ 0 & M & -MI_{23}I_{33}^{-1} \\ 0 & -I_{33}^{-1}I_{23}'M & I_{33}^{-1}+I_{33}^{-1}I_{23}'MI_{23}I_{33}^{-1} \end{bmatrix}$$

where $M = (I_{22}-I_{23}I_{33}^{-1}I_{23}')^{-1}$, so that

$$I^{-1}-\tilde{I} = \begin{bmatrix} 0 & 0 & 0 \\ 0 & M & -MI_{23}I_{33}^{-1} \\ 0 & -I_{33}^{-1}I_{23}^{'}M & I_{33}^{-1}I_{23}^{'}MI_{23}I_{33}^{-1} \end{bmatrix} = QMQ' ,$$

where $Q' = (0 \quad I_{n(n+1)/2} \quad -I_{23}I_{33}^{-1})$. However, $N^{\frac{1}{2}} \dfrac{\partial \tilde{\ell}_N(\theta_o)}{\partial \theta}$ is asymptotically distributed normally with mean zero and covariance matrix J which has been shown to equal $2I$ if $\{\varepsilon(t)\}$ and $\{B(t)\}$ are normal to the fourth order, a fact which enables us to obtain the following result.

THEOREM 6.1 Under the conditions (i)-(ix) and the further condition that $J = 2I$, both τ_N and $\tilde{\tau}_N$ are asymptotically distributed as χ^2 with $n(n+1)/2$ degrees of freedom under the hypothesis that $\gamma = \gamma_o$.

Proof. We shall prove the result for $\tilde{\tau}_N$ since the result will then hold for τ_N by lemma 6.1. The result will follow from a corollary of the Fisher-Cochran theorem if $\frac{1}{4}(I^{-1}-\tilde{I})(2I)(I^{-1}-\tilde{I}) = \frac{1}{2}(I^{-1}-\tilde{I})$. Now,

(6.2.2) $$I(I^{-1}-\tilde{I}) = \mathrm{I}-I\tilde{I}$$

$$= \mathrm{I} - \begin{bmatrix} I_{11} & 0 & 0 \\ 0 & I_{22} & I_{23} \\ 0 & I_{23}^{'} & I_{33} \end{bmatrix} \begin{bmatrix} I_{11}^{-1} & 0 & 0 \\ 0 & 0 & 0 \\ 0 & 0 & I_{33}^{-1} \end{bmatrix}$$

$$= \mathrm{I} - \begin{bmatrix} \mathrm{I} & 0 & 0 \\ 0 & 0 & I_{23}I_{33}^{-1} \\ 0 & 0 & 1 \end{bmatrix}$$

$$= \begin{bmatrix} 0 & 0 & 0 \\ 0 & \mathrm{I} & -I_{23}I_{33}^{-1} \\ 0 & 0 & 0 \end{bmatrix}$$

where the matrices I stand for identities with the relevant dimensions.

Thus

$$(I^{-1}-\tilde{I})I(I^{-1}-\tilde{I}) = QMQ'\begin{bmatrix} 0 & 0 & 0 \\ 0 & I & -I_{23}I_{33}^{-1} \\ 0 & 0 & 0 \end{bmatrix}$$

$$= QMQ' = I^{-1}-\tilde{I} .$$

Hence τ_N and $\tilde{\tau}_N$ are asymptotically distributed as χ^2 with degrees of freedom $tr[I(I^{-1}-\tilde{I})] = n(n+1)/2$, the trace of the matrix obtained in the final expression in (6.2.2). #

Although the quantity $\tilde{\tau}_N$ is not a statistic, it will be seen later that there is a statistic, depending only on scores, which converges in probability to $\tilde{\tau}_N$ and which therefore has the same asymptotic distribution as $\tilde{\tau}_N$ under the null hypothesis. Moreover, the distributional properties of $\tilde{\tau}_N$ are easily analysed even when the true parameter γ_o is such that $[\beta_o',r_o']'$ lies on the boundary of Θ, and in particular when $\gamma_o = 0$, as is illustrated in the following lemma.

LEMMA 6.2 <u>Under the conditions</u> (i)-(viii) <u>and the conditions that</u> $\sigma_o^2 \geq \delta_1 > 0$ <u>and</u> $\gamma_o = 0$, $\tilde{\tau}_N/\nu$ <u>is distributed asymptotically as</u> χ^2 <u>with</u> $n(n+1)/2$ <u>degrees of freedom, where</u>

$$\nu = \tfrac{1}{2}E[(\varepsilon^2(t)/\sigma_o^2)-1]^2 .$$

<u>Proof.</u> See appendix 6.1.

It is now possible to commence the derivation of a score test for the hypothesis that $\gamma = 0$. Let $\tilde{\beta}_N$ and $\tilde{\sigma}_N^2$ be the maximum likelihood estimates of β_o and σ_o^2 under that hypothesis. Then $\tilde{\beta}_N$ and $\tilde{\sigma}_N^2$ are the usual maximum likelihood estimates for a fixed coefficient autoregression, which are easily shown to be strongly consistent. We shall show that, if

$F_N(\beta,\sigma^2) = N^{\frac{1}{2}}\{I-\mathcal{I}\mathcal{I}\}\dfrac{\partial\tilde{\ell}_N(\beta,0,\sigma^2)}{\partial\theta}$, then $F_N(\beta_o,\sigma_o^2) - F_N(\tilde{\beta}_N,\tilde{\sigma}_N^2)$ converges in probability to zero, and hence that a test may be based on the vector of scores $\dfrac{\partial\tilde{\ell}_N(\tilde{\beta}_N,0,\tilde{\sigma}_N^2)}{\partial\theta}$. In order to do this we require the following.

LEMMA 6.3 Under the conditions of lemma 6.2, $F_N(\beta_o,\sigma_o^2) - F_N(\tilde{\beta}_N,\tilde{\sigma}_N^2)$ converges in probability to zero.

Proof. See appendix 6.1.

This lemma enables a formulation of a test for $\gamma = 0$ based on $F_N(\tilde{\beta}_N,\tilde{\sigma}_N^2)$. It was shown in lemma 6.2 that $\tilde{\tau}_N/\nu$ is asymptotically distributed as χ^2 with $n(n+1)/2$ degrees of freedom, where

$$\tilde{\tau}_N = \tfrac{1}{2}F_N'(\beta_o,\sigma_o^2)\mathcal{I}^{-1}F_N(\beta_o,\sigma_o^2) = \tfrac{1}{2}f_N'(\beta_o,\sigma_o^2)\,M\,f_N(\beta_o,\sigma_o^2)\ ,$$

$$f_N(\beta,\sigma^2) = N^{\frac{1}{2}}\left\{\frac{\partial\tilde{\ell}_N(\beta,0,\sigma^2)}{\partial\gamma} - I_{23}I_{33}^{-1}\frac{\partial\tilde{\ell}_N(\beta,0,\sigma^2)}{\partial\sigma^2}\right\}$$

and $M = (\mathcal{I}_{22}-\mathcal{I}_{23}\mathcal{I}_{33}^{-1}\mathcal{I}_{23}')^{-1}$. Let W_N, g_N, ν_N and $\hat{\tau}_N$ be defined by

$$W_N = N^{-1}\sum_{t=1}^{N}(z(t)-\bar{z})(z(t)-\bar{z})', \qquad g_N = N^{-1}\sum_{t=1}^{N}\tilde{\varepsilon}^2(t)z(t) - \tilde{\sigma}_N^2\bar{z}\ ,$$

$$\nu_N = \tfrac{1}{2}N^{-1}\sum_{t=1}^{N}[(\tilde{\varepsilon}^2(t)/\tilde{\sigma}_N^2)-1]^2$$

and $\hat{\tau}_N = (2\nu_N\tilde{\sigma}_N^4)^{-1}N\,g_N'W_N^{-1}g_N$, where $\tilde{\varepsilon}(t) = X(t) - \tilde{\beta}_N'Y(t-1)$

Then a test for $\gamma = 0$ may be based on the statistic $\hat{\tau}_N$, whose distributional properties under the hypothesis that $\gamma = 0$ are determined in the following theorem.

THEOREM 6.2 Under the conditions of lemma 6.2, $\hat{\tau}_N$ is asymptotically distributed as χ^2 with $n(n+1)/2$ degrees of freedom.

Proof. From the definition of I_{ij}, i,j = 2,3 it is seen that

$$I_{22}-I_{23}I_{33}^{-1}I_{23}' = \sigma_o^{-4}[E(z(t)z'(t))-E\{z(t)\}E\{z(t)\}'] .$$

Thus

$$\tilde{\tau}_N/\nu = \tfrac{1}{2}\nu^{-1}\sigma_o^{-4}f_N'(\beta_o,\sigma_o^2)[E(z(t)z'(t))-E\{z(t)\}E\{z(t)\}']^{-1}f_N(\beta_o,\sigma_o^2).$$

However, lemma 6.3 shows that $f_N(\beta_o,\sigma_o^2)$ may be replaced by $f_N(\tilde{\beta}_N,\tilde{\sigma}_N^2)$ $= -N^{-\frac{1}{2}}\tilde{\sigma}_N^{-4}\sum_{t=1}^{N}(\tilde{\varepsilon}^2(t)-\tilde{\sigma}_N^2)z(t)$. Hence, since $\tilde{\sigma}_N^2$ and W_N converge almost surely to σ_o^2 and $[E(z(t)z'(t))-E\{z(t)\}E\{z(t)\}']$ respectively and $g_N = -N^{-\frac{1}{2}}\tilde{\sigma}_N^{-4}f_N(\tilde{\beta}_N,\tilde{\sigma}_N^2)$ it is easily seen that $\hat{\tau}_N-\tilde{\tau}_N/\nu \overset{p}{\to} 0$ and that the result of the theorem follows. #

It should be noted here that the test proposed will not be equivalent to the likelihood ratio test, since the proof of lemma 6.1 is strongly dependent on the assumption that $[\beta_o',r_o']'$ is interior to Θ. The proof breaks down because the assumption that $\frac{\partial}{\partial\theta}\tilde{\ell}_N(\hat{\theta}_N) = 0$ is not necessarily true, $\ell_N(\beta,r)$ possibly being maximised on the boundary of Θ. Moreoever, if the assumption that $\frac{\partial}{\partial\theta}\tilde{\ell}_N(\hat{\theta}_N) = 0$ were correct, there would be a standard central limit theorem for $\hat{\theta}_N$, which is impossible since $[\beta_o',r_o']'$ is on the boundary of Θ. Consequently our test would be expected to be less powerful than a test based on the likelihood ratio statistic τ_N. However, the asymptotic distribution of the likelihood ratio test statistic τ_N may prove impossible to determine. Indeed, in a simpler instance, Moran (1971) and Chant (1974) showed how complicated it was to determine the distribution of maximum likelihood estimates when the true parameter lay on the boundary of the parameter space. In the case in question here, one would expect a non-zero probability that $[\hat{\beta}_N',\hat{r}_N']'$ lay on the boundary of Θ, with a non-zero probability that $\hat{r}_N$ equals zero.

6.3 An Alternative Test

In an attempt to recover some of the power lost by the test based on $\hat{\tau}_N$, we now consider a test statistic which reflects more the behaviour of the maximum likelihood estimate $\hat{\theta}_N$ and thus of the statistic τ_N.

It was shown in chapter 4 that if $[\beta_o', r_o']'$ is an interior point of Θ then

$$N^{\frac{1}{2}}\left[(\hat{\theta}_N-\theta_o)+I^{-1}\frac{\partial\tilde{\ell}_N(\theta_o)}{\partial\theta}\right] \overset{p}{\to} 0.$$

Thus

$$N^{\frac{1}{2}}\left[(\hat{\gamma}_N-\gamma_o)+(I_{22}-I_{23}I_{33}^{-1}I_{23}')^{-1}\left\{\frac{\partial\tilde{\ell}_N(\theta_o)}{\partial\gamma}-I_{23}I_{33}^{-1}\frac{\partial\tilde{\ell}_N(\theta_o)}{\partial\sigma^2}\right\}\right] \overset{p}{\to} 0.$$

Let $\gamma_N = W_N^{-1}g_N$, which is just the least-squares estimate of γ, and let Σ_N be the $n\times n$ symmetric matrix such that $\gamma_N = \text{vech}\,\Sigma_N$. Then from theorem 6.2, it is easily seen that even if $[\beta_o', r_o']'$ is not interior to Θ,

$$[N^{\frac{1}{2}}(\gamma_N-\gamma_o)+(I_{22}-I_{23}I_{33}^{-1}I_{23}')^{-1}f_N(\beta_o,\sigma_o^2)] \overset{p}{\to} 0. \tag{6.3.1}$$

An alternative test for $\gamma = 0$ is based on the statistic $S_N = N\,\tilde{\gamma}_N'W_N\tilde{\gamma}_N/(2\nu_N\tilde{\sigma}_N^4)$ where

$$\tilde{\gamma}_N = \begin{cases}\gamma_N & \text{if } \Sigma_N \text{ is non-negative definite}\\ 0 & \text{otherwise}.\end{cases}$$

The statistic $\tilde{\gamma}_N$ is designed both to partially correct the fact that $N^{\frac{1}{2}}(\hat{\gamma}_N-\gamma_o)+(I_{22}-I_{23}I_{33}^{-1}I_{23}')^{-1}f_N(\beta_o,\sigma_o^2)$ does not converge in probability to zero if $\gamma_o = 0$, as well as to use the information provided by (6.3.1).

Some preliminary results are required before the asymptotic distribution of S_N may be obtained.

LEMMA 6.4 Let Ω be a symmetric $n \times n$ random matrix with $\{\Omega_{ij};\ 1 \leq j \leq i \leq n\}$ an independent set of zero mean normal random variables and $E(\Omega_{ii}^2) = 1$, $E(\Omega_{ij}^2) = \frac{1}{2}$, $1 \leq j < i \leq n$. If $p_n = \Pr\{\Omega \text{ is non-negative definite}\}$ then

$$p_n = c \int_A \exp\left(-\tfrac{1}{2} \sum_{j=1}^{n} x_j^2\right) \prod_{j>k} (x_j - x_k)\, dx_1 \ldots dx_n$$

where $c^{-1} = (n!)^{-1}\, 2^{3n/2} \prod_{j=1}^{n} \Gamma(1 + \tfrac{1}{2}j)$ and $A \subset \mathbb{R}^n$ is defined by

$$A = \{(x_1, \ldots, x_n) : 0 < x_1 < x_2 \ldots < x_n < \infty\}.$$

Proof. See appendix 6.1.

In practice, the most commonly used orders would be less than five, for which $p_1 = \frac{1}{2}$, $p_2 = \frac{1}{2} - 2^{-3/2}$, $p_3 = \frac{1}{4} - 2^{-\frac{1}{2}}\pi^{-1}$, $p_4 = \frac{1}{4} - 2^{-7/2} - \frac{1}{2}\pi^{-1}$. These values, and a general formula for p_n when n is even, are derived in appendix 6.1.

LEMMA 6.5

1. If A is an $n \times n$ matrix, then $(A \otimes A)K_n' = (K_n' H_n)(A \otimes A)K_n'$.
2. If A is an $n \times n$ invertible matrix then $(H_n A \otimes A K_n')^{-1} = H_n(A^{-1} \otimes A^{-1})K_n'$ where H_n and K_n are defined in theorem A.1.3.

Proof. See appendix 6.1.

We now derive the asymptotic distribution of the statistic S_N.

THEOREM 6.3 Under the conditions of lemma 6.2,

$$\lim_{N \to \infty} \Pr\{S_N \leq x\} = 1 - \tilde{p}_n + \tilde{p}_n \Pr(X \leq x)$$

where X is distributed as χ^2 with $n(n+1)/2$ degrees of freedom, $\tilde{p}_n = \lim_{N \to \infty} \Pr\{\Sigma_N \geq 0\}$ and Σ_N is defined above (6.3.1). If $\{\varepsilon(t)\}$ is distributed normally to the fourth order, then $\tilde{p}_n = p_n$, defined in lemma 6.4.

Proof. From the definition of S_N,

$$\Pr\{S_N \le x\} = \Pr\{S_N \le x,\ \tilde{\gamma}_N = 0\} + \Pr\{S_N \le x,\ \tilde{\gamma}_N \ne 0\}$$

$$= \Pr\{\tilde{\gamma}_N = 0\} + \Pr\{N\gamma_N' W_N \gamma_N/(2\nu_N\tilde{\sigma}_N^4) \le x,\ \Sigma_N \ge 0\}$$

$$= 1 - \Pr\{\Sigma_N \ge 0\} + \Pr\{N\gamma_N' W_N \gamma_N/(2\nu_N\tilde{\sigma}_N^4) \le x,\ \Sigma_N \ge 0\}\ .$$

Let $Z = N^{\frac{1}{2}} A^{\frac{1}{2}} \gamma_N$ where $A^{\frac{1}{2}} = \lim_{N\to\infty} (2\nu_N\tilde{\sigma}_N^4)^{-\frac{1}{2}} W_N^{\frac{1}{2}} = \{E(\varepsilon^2(t) - \sigma_o^2)\}^{-\frac{1}{2}}$ $\{E[z(t)z'(t)] - E[z(t)]E[z'(t)]\}^{\frac{1}{2}}$. It has already been shown that Z is asymptotically normally distributed with mean zero and covariance matrix $I_{n(n+1)/2}$. Thus $Z = R\,f(\xi)$ where $R = (Z'Z)^{\frac{1}{2}}$, $f : \mathbb{R}^{n(n+1)/2-1} \to \mathbb{R}^{n(n+1)/2}$ and R and ξ are asymptotically independent ($R\,f(\xi)$ is the spherical polar representation of Z). But $\gamma_N = N^{-\frac{1}{2}} R A^{-\frac{1}{2}} f(\xi)$, and since the event $\{z'\Sigma_N z \ge 0,\ \forall z \in \mathbb{R}^n\}$ is equivalent to the event $\{z'(\Sigma_N/R)z \ge 0, \forall z \in \mathbb{R}^n\}$, the events $\{N\gamma_N' W_N \gamma_N/(2\nu_N\tilde{\sigma}_N^4) \le x\}$ and $\{\Sigma_N \ge 0\}$ are asymptotically independent since they depend only on R and ξ respectively. Consequently

$$\text{(6.3.2)}\quad \lim_{N\to\infty} \Pr\{S_N \le x\} = 1 - \tilde{p}_n + \lim_{N\to\infty} \Pr\{N\gamma_N' W_N \gamma_n/(2\nu_N\tilde{\sigma}_N^4) \le x,\ \Sigma_N \ge 0\}$$

$$= 1 - \tilde{p}_n + \lim_{N\to\infty} \Pr\{N\gamma_N' W_N \gamma_N/(2\nu_N\tilde{\sigma}_N^4) \le x\}\ .\ \tilde{p}_n$$

$$= 1 - \tilde{p}_n + \tilde{p}_n \Pr\{X \le x\}$$

where X is distributed as χ^2 with $n(n+1)/2$ degrees of freedom, since $\Pr\{N\gamma_N' W_N \gamma_N/(2\nu_N\tilde{\sigma}_N^4) \le x\} = \Pr\{\hat{\tau}_N \le x\}$. It thus remains to be shown that $\tilde{p}_n = p_n$ when $\{\varepsilon(t)\}$ is distributed normally to the fourth order. Now

$$\lim_{N\to\infty} \Pr\{\Sigma_N \ge 0\} = \lim_{N\to\infty} \Pr\{N^{\frac{1}{2}} \Sigma_N \ge 0\}$$

$$= \Pr\{\tilde{\Omega} \ge 0\}$$

where $\tilde{\Omega}$ is $n \times n$ and symmetric and $\operatorname{vech}\tilde{\Omega}$ is distributed normally with mean zero and covariance matrix A^{-1}, since $N^{\frac{1}{2}}\gamma_N$ has that asymptotic distribution. But $\tilde{\Omega}$ is non-negative definite if and only if $B\tilde{\Omega}B'$ is non-negative definite for some fixed $n \times n$ matrix B. Furthermore, $\mathrm{vech}(B\tilde{\Omega}B') = H_n(B \otimes B)K_n' \, \mathrm{vech}\,\tilde{\Omega}$ is distributed normally with mean zero and covariance matrix

$$H_n(B \otimes B)K_n' A^{-1} K_n(B' \otimes B')H_n'$$

whose inverse, if B is invertible, from lemma 6.5 is given by

$$(K_n((B^{-1})' \otimes (B^{-1})')H_n') \, A \, (H_n(B^{-1} \otimes B^{-1})K_n').$$

But $A = c^2 E\{(z(t)-E[z(t)])(z(t)-E[z(t)])'\}$ where $c^2 = \{E(\varepsilon^2(t)-\sigma_o^2)^2\}^{-1}$, and

$$\begin{aligned} z(t)-E[z(t)] &= K_n \, \mathrm{vec}[Y(t-1)Y'(t-1)-V] \\ &= K_n \, \mathrm{vec}[V^{\frac{1}{2}}(w(t-1)w'(t-1)-I_n)V^{\frac{1}{2}}] \\ &= K_n(V^{\frac{1}{2}} \otimes V^{\frac{1}{2}})K_n' \, \mathrm{vech}(w(t-1)w'(t-1)-I_n) \end{aligned}$$

where $V = E[Y(t-1)Y'(t-1)]$ and $w(t-1) = V^{-\frac{1}{2}}Y(t-1)$. However, if $\{\varepsilon(t)\}$ is normal to the fourth order, $Y(t-1)$ and $w(t-1)$ will be also, so that $w(t-1)$ is, to the fourth order, distributed normally with mean zero and covariance matrix I_n. If $B = V^{\frac{1}{2}}$ then from lemma 6.5 it follows that

$$\begin{aligned} &K_n((B^{-1})' \otimes (B^{-1})')H_n'\{z(t)-E[z(t)]\} \\ &= K_n(V^{-\frac{1}{2}} \otimes V^{-\frac{1}{2}})H_n' K_n(V^{\frac{1}{2}} \otimes V^{\frac{1}{2}})K_n' \, \mathrm{vech}\{w(t-1)w'(t-1)-I_n\} \\ &= K_n(V^{-\frac{1}{2}} \otimes V^{-\frac{1}{2}})H_n' K_n(V^{\frac{1}{2}} \otimes V^{\frac{1}{2}})H_n' K_n K_n' \, \mathrm{vech}\{w(t-1)w'(t-1)-I_n\} \\ &= (K_n K_n') \, \mathrm{vech}\{w(t-1)w'(t-1)-I_n\}. \end{aligned}$$

Letting $w(t-1) = [w_1,\ldots,w_n]'$, the elements of the covariance matrix of $\mathrm{vech}\{w(t-1)w'(t-1)-I_n\}$ are seen to be of the form $E(w_i w_j - \delta_{ij})(w_k w_\ell - \delta_{k\ell})$ $i \le j$ and $k \le \ell$, where δ_{ij} is Kronecker's delta. Thus the only non-zero elements of the covariance matrix occur when $i = j = k = \ell$, and $i = k \neq j = \ell$. When $i = j = k = \ell$, $E(w_i w_j - \delta_{ij})(w_k w_\ell - \delta_{k\ell}) = E(w_i^2 - 1)^2 = 2$ and when $i = k \neq j = \ell$, $E(w_i w_j - \delta_{ij})(w_k w_\ell - \delta_{k\ell}) = E(w_i^2 w_k^2) = 1$.

Let Z be any $n \times n$ symmetric matrix. Then

$$(\text{vec}\, Z)'(\text{vec}\, Z) = (\text{vech}\, Z)'(K_n K_n')\, \text{vech}\, Z.$$

But

$$(\text{vec}\, Z)'(\text{vec}\, Z) = \sum_{i,j=1}^{n} z_{ij}^2$$

$$= \sum_{i=1}^{n} z_{ii}^2 + 2 \sum_{1 \leq i < j \leq n} z_{ij}^2$$

$$= (\text{vech}\, Z)'\, M\, (\text{vech}\, Z)$$

where M is diagonal with diagonal elements 1 and 2, and equals $K_n K_n'$. By comparison with the covariance matrix of $\text{vech}\{w(t-1)w'(t-1) - I_n\}$ it is thus seen that this covariance matrix equals $2(K_n K_n')^{-1}$, and that the covariance matrix of $K_n(V^{-\frac{1}{2}} \otimes V^{-\frac{1}{2}}) H_n' \{z(t) - E[z(t)]\}$ is $2 K_n K_n'$. Hence $\text{vech}(V^{\frac{1}{2}} \tilde{\Omega} V^{\frac{1}{2}})$ is distributed normally with mean zero and covariance matrix $(2c^2 K_n K_n')^{-1}$, and

$$\Pr\{\tilde{\Omega} \geq 0\} = \Pr\{\sqrt{2}\, c\, V^{\frac{1}{2}} \tilde{\Omega} V^{\frac{1}{2}} \geq 0\}$$

$$= \Pr\{\Omega \geq 0\}$$

where $\text{vech}\, \Omega$ is distributed normally with mean zero and covariance matrix $(K_n K_n')^{-1}$. Thus Ω satisfies the conditions of lemma 6.4 and $\tilde{p}_n = p_n$. #

6.4 Power Comparisons

In the previous two sections, the asymptotic distributions of the two statistics $\hat{\tau}_N$ and S_N were obtained under the null hypothesis. It was also stated in §6.3 that the power of the test based on S_N would be better than the test based on $\hat{\tau}_N$. Because of the relationship between $\hat{\tau}_N$ and S_N, it is a simple matter to justify this claim. The size-α critical regions for rejecting $H_0\colon \Sigma = 0$ are, by theorems 6.2 and 6.3,

$$\{\hat{\tau}_N : \hat{\tau}_N > \chi^2_{n(n+1)/2}(\alpha)\}$$

and

$$\{S_N : S_N > \chi^2_{n(n+1)/2}(\alpha/\tilde{p}_n)\}$$

where $\Pr\{X > \chi^2_{n(n+1)/2}(\alpha)\} = \alpha$ if X is distributed as χ^2 with $n(n+1)/2$ degrees of freedom.

THEOREM 6.4 <u>Under the sequence of hypotheses</u> $\{H_N : \gamma = N^{-\frac{1}{2}}\gamma^*\}$ <u>where</u> $\gamma^* = \text{vech}(\Sigma^*)$ <u>and</u> Σ^* <u>is symmetric and positive definite, the test based on</u> S_N <u>is asymptotically better than that based on</u> $\hat{\tau}_N$ <u>in that the hypothesis</u> H_o <u>that</u> $\Sigma = 0$ <u>is rejected by the former test with higher probability</u>.

<u>Proof</u>. Theorem 3.2 shows that under $\{H_N\}$, $N^{\frac{1}{2}}\gamma_N$ converges in distribution to the normal with mean γ^* and covariance matrix Ω_{22}, where Ω_{22} here will be the covariance matrix of $N^{\frac{1}{2}}\gamma_N$ under H_o, or the matrix A^{-1} defined in theorem 6.3. Consequently $N^{\frac{1}{2}}(\gamma_N - \tilde{\gamma}_N)$ converges in probability to zero since $\lim_{N\to\infty} \Pr\{\tilde{\gamma}_N = 0 | H_N\} = 0$, and $(S_N - \hat{\tau}_N) \xrightarrow{p} 0$. Hence, under $\{H_n\}$, the asymptotic probability that the test based on S_N rejects H_o is given by

$$\lim_{N\to\infty} \Pr\{S_N > \chi^2_{n(n+1)/2}(\alpha/\tilde{p}_n) | H_N\}$$

$$= \lim_{N\to\infty} \Pr\{\hat{\tau}_N > \chi^2_{n(n+1)/2}(\alpha/\tilde{p}_n) | H_N\}$$

$$> \lim_{N\to\infty} \Pr\{\hat{\tau}_N > \chi^2_{n(n+1)/2}(\alpha) | H_N\}$$

since $\tilde{p}_n < 1$.

Moreover, since $N^{\frac{1}{2}}A^{\frac{1}{2}}\gamma_N$ is asymptotically normally distributed with mean $A^{\frac{1}{2}}\gamma^*$ and covariance matrix $I_{n(n+1)/2}$ under $\{H_N\}$, $\hat{\tau}_N$ will be distributed asymptotically as non-central χ^2 with $n(n+1)/2$ degrees of freedom and non-centrality parameter $(\gamma^*)'A\gamma^*$ under $\{H_N\}$. Hence the asymptotic difference between the power functions is given by

$$\Pr\{\chi^2_{n(n+1)/2}(\alpha/\tilde{p}_n) < X \leq \chi^2_{n(n+1)/2}(\alpha)\}$$

where X is distributed as non-central χ^2 with $n(n+1)/2$ degrees of freedom and non-centrality parameter $(\gamma^*)'A\gamma^*$. #

It should be noted that the above theorem does not solve the question of power with respect to the alternative sequence of hypotheses $\{H_N : \gamma = N^{-\frac{1}{2}}\gamma^*\}$ if $\gamma^* = \text{vech}(\Sigma^*)$ where Σ^* is *not* positive definite. While the question is a relevant one, the alternative hypotheses dealt with in theorem 6.4 would usually be of more interest.

APPENDIX 6.1

PROOF OF LEMMA 6.1 The proof follows essentially by Taylor series expansions of $\tilde{\ell}_N$ about θ_o. Now

$$\tilde{\ell}_N(\hat{\theta}_N) = \tilde{\ell}_N(\theta_o) + \frac{\partial \tilde{\ell}_N(\theta_o)}{\partial \theta'} (\hat{\theta}_N - \theta_o) + \tfrac{1}{2}(\hat{\theta}_N - \theta_o)' \frac{\partial^2 \tilde{\ell}_N(\theta_N^*)}{\partial \theta \partial \theta'} (\hat{\theta}_N - \theta_o) \tag{A.6.1}$$

where θ_N^* lies on the line segment between θ_o and $\hat{\theta}_N$ and consequently converges almost surely to θ_o by corollary 4.2.1. However, using the proof of theorem 4.3 it can be shown that

$$N^{\frac{1}{2}} \left[\frac{\partial \tilde{\ell}_N(\theta_o)}{\partial \theta} + \frac{\partial^2 \tilde{\ell}_N(\theta_o)}{\partial \theta \partial \theta'} (\hat{\theta}_N - \theta_o) \right] \overset{p}{\to} 0 .$$

Consequently, from (A.6.1),

$$N \left\{ \tilde{\ell}_N(\hat{\theta}_N) - \tilde{\ell}_N(\theta_o) + \tfrac{1}{2} \frac{\partial \tilde{\ell}_N(\theta_o)}{\partial \theta'} \left(\frac{\partial^2 \tilde{\ell}_N(\theta_o)}{\partial \theta \partial \theta'} \right)^{-1} \frac{\partial \tilde{\ell}_N(\theta_o)}{\partial \theta} \right\} \overset{p}{\to} 0 \tag{A.6.2}$$

since $\dfrac{\partial^2 \tilde{\ell}_N(\theta)}{\partial \theta \partial \theta'}$ is continuous in a neighbourhood of θ_o and $\theta_N^* \overset{p}{\to} \theta_o$. Similarly for $\tilde{\ell}_N(\tilde{\theta}_N)$ we obtain

$$\tilde{\ell}_N(\tilde{\theta}_N) = \tilde{\ell}_N(\theta_o) + d_N'(\tilde{\theta}_N - \theta_o) + \tfrac{1}{2}(\tilde{\theta}_N - \theta_o)' M_N(\theta_N^{**})(\tilde{\theta}_N - \theta_o) \tag{A.6.3}$$

where

$$d_N' = \left[\frac{\partial \tilde{\ell}_N(\theta_o)}{\partial \beta'} \quad 0 \quad \frac{\partial \tilde{\ell}_N(\theta_o)}{\partial \sigma^2} \right],$$

$$M_N(\theta) = \begin{bmatrix} \frac{\partial^2\tilde{\ell}_N(\theta)}{\partial\beta\partial\beta'} & 0 & \frac{\partial^2\tilde{\ell}_N(\theta)}{\partial\beta\partial\sigma^2} \\ 0 & 0 & 0 \\ \frac{\partial^2\tilde{\ell}_N(\theta)}{\partial\sigma^2\partial\beta'} & 0 & \frac{\partial^2\tilde{\ell}_N(\theta)}{\partial(\sigma^2)^2} \end{bmatrix},$$

d_N is $n(n+1)/2 \times 1$, $M_N(\theta)$ is $n(n+1)/2 \times n(n+1)/2$ and θ_N^{**} lies on the line segment between θ_o and $\tilde{\theta}_N$. A slight modification of the proof of theorem 4.3 shows that

$$N^{\frac{1}{2}}\{d_N+M_N(\theta_o)(\tilde{\theta}_N-\theta_o)\} \overset{P}{\to} 0$$

and we then have

$$\text{(A.6.4)} \qquad N\left\{\tilde{\ell}_N(\tilde{\theta}_N)-\tilde{\ell}_N(\theta_o)+\tfrac{1}{2}\frac{\partial\tilde{\ell}_N(\theta_o)}{\partial\theta'}P_N\frac{\partial\tilde{\ell}_N(\theta_o)}{\partial\theta}\right\} \overset{P}{\to} 0$$

where

$$P_N = \begin{bmatrix} P_{11} & 0 & P_{12} \\ 0 & 0 & 0 \\ P_{21} & 0 & P_{22} \end{bmatrix} \text{ and } \begin{bmatrix} P_{11} & P_{12} \\ P_{21} & P_{22} \end{bmatrix} = \begin{bmatrix} \frac{\partial^2\tilde{\ell}_N(\theta_o)}{\partial\beta\partial\beta'} & \frac{\partial^2\tilde{\ell}_N(\theta_o)}{\partial\beta\partial\sigma^2} \\ \frac{\partial^2\tilde{\ell}_N(\theta_o)}{\partial\sigma^2\partial\beta'} & \frac{\partial^2\tilde{\ell}_N(\theta_o)}{\partial(\sigma^2)^2} \end{bmatrix}^{-1}.$$

Combining (A.6.2) and (A.6.4) we obtain

$$N\left\{\tilde{\ell}_N(\hat{\theta}_N)-\tilde{\ell}_N(\tilde{\theta}_N)+\tfrac{1}{2}\frac{\partial\tilde{\ell}_N(\theta_o)}{\partial\theta'}\left[\left(\frac{\partial^2\tilde{\ell}_N(\theta_o)}{\partial\theta\partial\theta'}\right)^{-1}-P_N\right]\frac{\partial\tilde{\ell}_N(\theta_o)}{\partial\theta}\right\} \overset{P}{\to} 0$$

and noting that $\frac{\partial^2\tilde{\ell}_N(\theta_o)}{\partial\theta\partial\theta'} \overset{a.s.}{\to} I$, $I_{12} = 0$ and $I_{13} = 0$, the result given in the lemma follows. #

PROOF OF LEMMA 6.2 Neither the derivation of the matrix I nor the central limit theorem for $N^{\frac{1}{2}} \frac{\partial \tilde{\ell}_N(\theta_o)}{\partial \theta}$ depends on $[\beta_o', r_o']'$ being interior to Θ. Consequently these results remain valid when $\gamma_o = 0$. Moreover, an inspection of the matrices J and I reveals that $J_{ij} = 2\nu I_{ij}$; $i,j = 2,3$. Now, in the notation of theorem 5.1,

$$\tfrac{1}{4}(I^{-1}-\tilde{I})\, J\, (I^{-1}-\tilde{I}) = \tfrac{1}{4}I^{-1}(\mathrm{I}-I\tilde{I})\, J\, (\mathrm{I}-\tilde{I}I)I^{-1}.$$

But

$$(\mathrm{I}-I\tilde{I})\, J\, (\mathrm{I}-\tilde{I}I) = \begin{bmatrix} 0 & 0 & 0 \\ 0 & \mathrm{I} & -I_{23}I_{33}^{-1} \\ 0 & 0 & 0 \end{bmatrix} \begin{bmatrix} J_{11} & J_{12} & J_{13} \\ J_{12}' & J_{22} & J_{23} \\ J_{13}' & J_{23}' & J_{33} \end{bmatrix} \begin{bmatrix} 0 & 0 & 0 \\ 0 & \mathrm{I} & 0 \\ 0 & -I_{33}^{-1}I_{23}' & 0 \end{bmatrix}$$

$$= \begin{bmatrix} 0 & 0 & 0 \\ 0 & \mathrm{I} & -I_{23}I_{33}^{-1} \\ 0 & 0 & 0 \end{bmatrix} \begin{bmatrix} 2\nu I_{11} & 0 & 0 \\ 0 & 2\nu I_{22} & 2\nu I_{23} \\ 0 & 2\nu I_{23}' & 2\nu I_{33} \end{bmatrix} \begin{bmatrix} 0 & 0 & 0 \\ 0 & \mathrm{I} & 0 \\ 0 & -I_{23}'I_{33}^{-1} & 0 \end{bmatrix}$$

$$= 2\nu(\mathrm{I}-I\tilde{I})\, I\, (\mathrm{I}-\tilde{I}I).$$

Thus, from the proof of theorem 5.1,

$$\tfrac{1}{4}(I^{-1}-\tilde{I})\, J\, (I^{-1}-\tilde{I}) = \nu/2(I^{-1}-\tilde{I})$$

and $\tilde{\tau}_N/\nu$ is asymptotically distributed as χ^2 with $n(n+1)/2$ degrees of freedom. #

PROOF OF LEMMA 6.3 From the definition of $F_N(\beta,\sigma^2)$ we have

$$F_N'(\beta,\sigma^2) = N^{\frac{1}{2}}\left[0, \left(\frac{\partial\tilde{\ell}_N(\beta,0,\sigma^2)}{\partial\gamma} - I_{23}I_{33}^{-1}\frac{\partial\tilde{\ell}_N(\beta,0,\sigma^2)}{\partial\sigma^2}\right)', 0\right]$$

the two null vectors being of dimensions n and 1 respectively. In the notation of chapter 4 which defines the matrix I, we have, when $\gamma = 0$, $u_{ot} = \varepsilon(t)$, $\lambda_{ot} = \sigma_o^2$ and $\eta(t) = \varepsilon^2(t)-\sigma_o^2$. Thus $I_{23} = \sigma_o^{-4}E[z(t)]$ and $I_{33} = \sigma_o^{-4}$. Also,

$$\frac{\partial}{\partial\gamma}\tilde{\ell}_N(\beta,0,\sigma^2) = -\sigma^{-4}N^{-1}\sum_{t=1}^{N}[(X(t)-\beta'Y(t-1))^2-\sigma^2]z(t)$$

and

$$\frac{\partial}{\partial\sigma^2}\tilde{\ell}_N(\beta,0,\sigma^2) = -\sigma^{-4}N^{-1}\sum_{t=1}^{N}[(X(t)-\beta'Y(t-1))^2-\sigma^2] .$$

Hence, letting

$$f_N(\beta,\sigma^2) = N^{\frac{1}{2}}\left\{\frac{\partial\tilde{\ell}_N(\beta,0,\sigma^2)}{\partial\gamma} - I_{23}I_{33}^{-1}\frac{\partial\tilde{\ell}_N(\beta,0,\sigma^2)}{\partial\sigma^2}\right\}$$

we have

$$f_N(\beta,\sigma^2) = -\sigma^{-4}N^{-\frac{1}{2}}\sum_{t=1}^{N}[(X(t)-\beta'Y(t-1))^2-\sigma^2][z(t)-E\{z(t)\}]$$

and

$$f_N(\tilde{\beta}_N,\tilde{\sigma}_N^2)-f_N(\beta_o,\sigma_o^2) = N^{-\frac{1}{2}}\sigma_o^{-4}\sum_{t=1}^{N}(\varepsilon^2(t)-\sigma_o^2)(z(t)-E\{z(t)\})$$

$$-N^{-\frac{1}{2}}\tilde{\sigma}_N^{-4}\sum_{t=1}^{N}(\tilde{\varepsilon}^2(t)-\tilde{\sigma}_N^2)(z(t)-E\{z(t)\})$$

where $\tilde{\varepsilon}(t) = X(t)-\tilde{\beta}_N'Y(t-1) = \varepsilon(t)+(\beta_o-\tilde{\beta}_N)'Y(t-1)$. Therefore

$$f_N(\tilde\beta_N,\tilde\sigma_N^2)-f_N(\beta_o,\sigma_o^2) = N^{\frac{1}{2}}(\tilde\sigma_N^2\sigma_o^2)^{-1}(\sigma_o^2-\tilde\sigma_N^2)(\bar z-E\{z(t)\})$$

$$-N^{-\frac{1}{2}}(\tilde\sigma_N^{-4}-\sigma_o^{-4})\sum_{t=1}^{N}\varepsilon^2(t)(z(t)-E\{z(t)\})$$

$$-N^{-\frac{1}{2}}\tilde\sigma_N^{-4}\sum_{t=1}^{N}(\tilde\varepsilon^2(t)-\varepsilon^2(t))(z(t)-E\{z(t)\})$$

$$= A_1 - A_2 - A_3, \quad \text{say.}$$

Now, from the standard theory for fixed coefficient autoregressions it follows that $\tilde\beta_N$ and $\tilde\sigma_N^2$ converge almost surely to β_o and σ_o^2 respectively, and that $N^{\frac{1}{2}}(\tilde\beta_N-\beta_o)$ and $N^{\frac{1}{2}}(\tilde\sigma_N^2-\sigma_o^2)$ have distributions which converge to normal distributions with means zero. Thus it is easily seen that the terms A_1 and A_2 converge in probability to zero, by noting that $\bar z-E\{z(t)\}$ and $N^{-1}\sum_{t=1}^{N}\varepsilon^2(t)(z(t)-E\{z(t)\})$ converge almost surely to zero by the ergodic theorem. The term A_3 is seen to converge in probability to zero by the following argument:

$$\sum_{t=1}^{N}(\tilde\varepsilon^2(t)-\varepsilon^2(t))(z(t)-E\{z(t)\}) = \sum_{t=1}^{N}(\tilde\varepsilon(t)-\varepsilon(t))(\tilde\varepsilon(t)+\varepsilon(t))(z(t)-E\{z(t)\})$$

$$= \sum_{t=1}^{N}[2\varepsilon(t)+(\beta_o-\tilde\beta_N)'Y(t-1)][z(t)-E\{z(t)\}]Y'(t-1)(\beta_o-\tilde\beta_N)$$

$$= 2\sum_{t=1}^{N}\varepsilon(t)[z(t)-E\{z(t)\}]Y'(t-1)(\beta_o-\tilde\beta_N) + \sum_{t=1}^{N}[(\beta_o-\tilde\beta_N)'Y(t-1)]^2[z(t)-E\{z(t)\}].$$

But $N^{-1}\sum_{t=1}^{N}\varepsilon(t)[z(t)-E\{z(t)\}]Y'(t-1)$ and $N^{-1}\sum_{t=1}^{N}z(t)(Y'(t-1)\otimes Y'(t-1))$ converge almost surely to zero and $E\{z(t)(Y'(t-1)\otimes Y'(t-1))\}$ respectively by the ergodic theorem and $N^{\frac{1}{2}}(\beta_o-\tilde\beta_N)$ converges to a normal distribution with mean zero.
Thus $N^{-\frac{1}{2}}\sum_{t=1}^{N}(\tilde\varepsilon^2(t)-\varepsilon^2(t))(z(t)-E\{z(t)\}) \xrightarrow{p} 0$, as does A_3 since $\tilde\sigma_N^2 \xrightarrow{a.s.} \sigma_o^2$.

#

PROOF OF LEMMA 6.4 The joint density of the set of random variables $\{\Omega_{ij}, 1 \le j \le i \le n\}$ is

$$(2\pi)^{-n(n+1)/2}\ 2^{n(n-1)/2}\ \exp\{-\tfrac{1}{2}\,\mathrm{tr}(\Omega^2)\}\ .$$

The theory of random matrices composed of elements with such joint densities has been considered extensively in the theoretical physics literature. The random matrix Ω, above, is said to represent the Hamiltonian of a system in the Gaussian orthogonal ensemble. The eigenvalues of Ω are used to model the local statistical behaviour of the energy levels of certain ideal types of nuclei, and there are a number of methods available for deriving the properties of these eigenvalues, these methods being described in a unified fashion by Mehta (1967).

It may be shown (Mehta pp.31-33) that the joint density of the eigenvalues $\{x_1,\dots,x_n\}$ of Ω is

$$k_n \exp(-\tfrac{1}{2}\sum_{j=1}^{n} x_j^2) \prod_{i>j} |x_i - x_j|$$

where $k_n^{-1} = 2^{3n/2} \prod_{j=1}^{n} \Gamma(1+\tfrac{1}{2}j)$. Thus

$$p_n = k_n \int_{\substack{x_j \ge 0 \\ j=1,..,n}} \exp(-\tfrac{1}{2}\sum_{j=1}^{n} x_j^2) \prod_{j>k} |x_j - x_k|\ dx_1,\dots,dx_n$$

$$= n!\,k_n \int_A \exp(-\tfrac{1}{2}\sum_{j=1}^{n} x_j^2) \prod_{j>k} (x_j - x_k)\ dx_1,\dots,dx_n$$

where $A \subset \mathbb{R}^n$ is defined as $\{(x_1,\dots,x_n) : 0 < x_1 < x_2 \dots < x_n < \infty\}$, the last expression for p_n following because of the symmetry of the integrand. #

The Evaluation of p_n

In order to evaluate p_n, let

$$\phi_j(x) = c_j^{-\frac{1}{2}} \exp(-\tfrac{1}{2}x^2) H_j(x),$$

where $c_j = \pi^{\frac{1}{2}} 2^j j!$ and $H_j(x)$ is the Hermite polynomial of order j. Then, following Mehta p.51, it can be seen that

$$\text{(A.6.5)} \qquad p_n = n!\, k_n\, 2^{-n(n-1)/2} \prod_{j=o}^{n-1} c_j^{\frac{1}{2}} \int_A \det[\Lambda(x_1,\ldots,x_n)]\, dx_1,\ldots,dx_n,$$

where Λ is the $n \times n$ matrix whose ij^{th} component is $\phi_{i-1}(x_j)$. If the odd-indexed variables $x_1, x_3, \ldots$ are integrated out, the remaining integrand will be symmetric in the even-indexed variables, facilitating evaluation of p_n. The effect of integrating out $x_1, x_3, \ldots$ is to replace the odd-numbered columns of Λ with columns whose i^{th} elements are $\int_o^{x_{j+1}} \phi_{i-1}(x)\,dx$, with x_{n+1} replaced by ∞ if n is odd. Thus

$$\text{(A.6.6)} \qquad p_n = n!\, k_n \left(\prod_{j=o}^{n-1} c_j^{\frac{1}{2}} \right) 2^{-n(n-1)/2} \int_B \det\{\tilde{\Lambda}(x_2,\ldots,x_{2[n/2]})\}\, dx_2,\ldots,dx_{2[n/2]}$$

where $B \subset \mathbb{R}^{[n/2]}$ is defined by $B = \{(x_2, x_4, \ldots, x_{2[n/2]}) : 0 < x_2 < x_4 < \ldots < x_{2[n/2]} < \infty\}$ and

$$\tilde{\Lambda}_{ij} = \begin{cases} \phi_{i-1}(x_j) & j = 2,4,\ldots,2[n/2] \\ \int_o^{x_{j+1}} \phi_{i-1}(x)\,dx & j = 1,3,\ldots,2[n/2]-1 \\ \int_o^{\infty} \phi_{i-1}(x)\,dx & n \text{ odd and } j = n. \end{cases}$$

A further simplification is possible if n is even. Letting $n = 2m$, it may be shown that

$$\text{(A.6.7)} \qquad p_n = (\det M)^{\frac{1}{2}}$$

where M is the $2m \times 2m$ skew-symmetric matrix whose ij^{th} 2×2 submatrix M_{ij} is defined by

$$M_{ij} = \begin{bmatrix} \lambda_{i-1,j-1} & g_{i-1,j-1} \\ -g_{j-1,i-1} & \mu_{i-1,j-1} \end{bmatrix}$$

$$\lambda_{ij} = \int_0^\infty \int_y^\infty \{\phi_{2i}(y)\phi_{2j}(x) - \phi_{2i}(x)\phi_{2j}(y)\}\, dx\, dy$$

$$\mu_{ij} = \tfrac{1}{4} \int_0^\infty \int_y^\infty \{\phi'_{2i}(y)\phi'_{2j}(x) - \phi'_{2i}(x)\phi'_{2j}(y)\}\, dx\, dy$$

$$g_{ij} = -\tfrac{1}{2} \int_0^\infty \int_y^\infty \{\phi_{2i}(y)\phi'_{2j}(x) - \phi_{2i}(x)\phi'_{2j}(y)\}\, dx\, dy.$$

Furthermore, since M is skew-symmetric, det M may be expressed as the square of a certain sum of products of the upper-triangular elements of M (for details see Mehta pp.51-53, 194-195), so that the evaluation of det M is unnecessary.

It is easy to see that $p_1 = \tfrac{1}{2}$. Also, from (A.6.7), it follows that

$$p_2 = g_{oo}$$

$$= \tfrac{1}{2}\int_0^\infty \phi_o^2(y)\, dy + \tfrac{1}{2}\int_0^\infty \phi_o^2(y)\, dy - \tfrac{1}{2}\left\{\int_0^\infty \phi_o(x)\, dx\right\}\phi_o(0)$$

$$p_2 = c_o^{-1}\int_0^\infty \exp(-y^2)\, dy - \tfrac{1}{2}c_o^{-1}\int_0^\infty \exp(-\tfrac{1}{2}y^2)\, dy$$

$$= \pi^{-\frac{1}{2}}\{\tfrac{1}{2}\pi^{\frac{1}{2}} - \tfrac{1}{4}(2\pi)^{\frac{1}{2}}\}$$

$$= \tfrac{1}{2} - 2^{-3/2}\ .$$

To evaluate p_3 we use (A.6.6). The matrix $\tilde{\Lambda}(x_2)$ has as its second column the vector $[\phi_o(x_2), \phi_1(x_2), \phi_2(x_2)]'$ where

$$\phi_o(x) = \pi^{-\frac14} \exp(-\tfrac12 x^2)$$

$$\phi_1(x) = 2^{-\frac12} \pi^{-\frac14} 2x \exp(-\tfrac12 x^2)$$

$$= 2^{\frac12} \pi^{-\frac14} x \exp(-\tfrac12 x^2)$$

and

$$\phi_2(x) = 2^{-3/2} \pi^{-\frac14} (4x^2-2) \exp(-\tfrac12 x^2)$$

$$= 2^{-\frac12} \pi^{-\frac14} (2x^2-1) \exp(-\tfrac12 x^2) .$$

The first column is thus $[f_o(x_2), f_1(x_2), f_2(x_2)]'$ where

$$f_o(x) = \pi^{-\frac14} \int_o^x \exp(-\tfrac12 u^2)\, du$$

$$= 2^{\frac12} \pi^{\frac14} \{\Phi(x) - \tfrac12\}$$

$$f_1(x) = 2^{\frac12} \pi^{-\frac14} \int_o^x u \exp(-\tfrac12 u^2)\, du$$

$$= 2^{\frac12} \pi^{-\frac14} \{1-\exp(-\tfrac12 x^2)\}$$

$$f_2(x) = 2^{-\frac12} \pi^{-\frac14} \int_o^x (2u^2-1) \exp(-\tfrac12 u^2)\, du$$

$$= 2^{-\frac12} \pi^{-\frac14} \{-2x \exp(-\tfrac12 x^2) + 2^{\frac12} \pi^{\frac12} (\Phi(x) - \tfrac12)\}$$

and

$$\Phi(x) = \int_{-\infty}^x (2\pi)^{-\frac12} \exp(-\tfrac12 u^2)\, du .$$

Hence the third column of $\tilde{\Lambda}(x_2)$ is the vector $[f_o(\infty), f_1(\infty), f_2(\infty)]'$ where

$$f_o(\infty) = 2^{-\frac12} \pi^{\frac14}$$

$$f_1(\infty) = 2^{\frac12} \pi^{-\frac14}$$

and

$$f_2(\infty) = 2^{-1} \pi^{\frac14} .$$

After subtracting $2^{-\frac12}$ times the first row from the third row $\det \tilde{\Lambda}(x)$ is seen to equal

$$\begin{vmatrix} 2^{\frac{1}{2}}\pi^{\frac{1}{4}}\{\Phi(x)-\frac{1}{2}\} & \pi^{-\frac{1}{4}}\exp(-\frac{1}{2}x^2) & 2^{-\frac{1}{2}}\pi^{\frac{1}{4}} \\ 2^{\frac{1}{2}}\pi^{-\frac{1}{4}}\{1-\exp(-\frac{1}{2}x^2)\} & 2^{\frac{1}{2}}\pi^{-\frac{1}{4}}x\exp(-\frac{1}{2}x^2) & 2^{\frac{1}{2}}\pi^{-\frac{1}{4}} \\ -2^{\frac{1}{2}}\pi^{-\frac{1}{4}}x\exp(-\frac{1}{2}x^2) & 2^{\frac{1}{2}}\pi^{-\frac{1}{4}}(x^2-1)\exp(-\frac{1}{2}x^2) & 0 \end{vmatrix}$$

whose integral over B may be shown to be $2^{-3/2}\pi^{\frac{1}{4}}-\pi^{-3/4}$. Thus

$$p_3 = 3!\,k_3\,2^{-2}\prod_{j=o}^{2} c_j^{\frac{1}{2}}(2^{-3/2}\pi^{\frac{1}{4}}-\pi^{-3/2})$$

$$= \tfrac{1}{4} - 2^{-\frac{1}{2}}\pi^{-1}\,.$$

In order to evaluate p_{2n} when n is an integer, in view of (A.6.7) one need only calculate the integrals λ_{ij}, μ_{ij} and g_{ij}. To facilitate evaluation of these integrals we shall need the following results.

(i) $\phi_{2i}(0) = (-1)^i\,\pi^{-\frac{1}{4}}d_i$ where $d_i = \{(2i)!\}^{\frac{1}{2}}/(2^i i!)$

and $\phi_{2i+1}(0) = 0$, $i = 0,1,2,\ldots$.

(ii) $\int_0^\infty \phi_{2i}(y)\,dy = 2^{-\frac{1}{2}}\pi^{\frac{1}{4}}d_i$, $i = 0,1,2,\ldots$.

(iii) $\int_0^\infty \phi_{2i}(y)\phi_{2j}(y)\,dy = \frac{1}{2}\delta_{ij}$, $i,j = 0,1,2,\ldots$ where δ_{ij} is Kronecker's delta.

(iv) $\phi_{2j}(x) = (2j)^{-\frac{1}{2}}\{(2j-1)^{\frac{1}{2}}\phi_{2j-2}(x) - 2^{\frac{1}{2}}\phi'_{2j-1}(x)\}$, $j \geq 1$

where $\phi'_k(x) = \frac{d}{dx}(\phi_k(x))$.

(v) $\int_0^\infty \phi_{2i}(y)\phi_{2j-1}(y)\,dy = (-1)^{i+j+1}\pi^{-\frac{1}{2}}j^{\frac{1}{2}}d_i d_j/(2j-1-2i)$.

The proofs of these results rely on the properties of the Hermite polynomials and, although tedious, are quite straightforward to verify and will not be reproduced here.

Now

$$\lambda_{ij} = 2\int_0^\infty \int_y^\infty \phi_{2i}(y)\phi_{2j}(x)\,dxdy - \int_0^\infty \phi_{2i}(x)\,dx \int_0^\infty \phi_{2i}(y)\,dy$$

$$= 2\alpha_{ij} - 2^{-1}\pi^{\frac{1}{2}} d_i d_j$$

where

$$\alpha_{ij} = \int_0^\infty \int_y^\infty \phi_{2i}(y)\phi_{2j}(x)\,dxdy\,.$$

From (iv) it follows that

$$\alpha_{ij} = \{(2j-1)/(2j)\}^{\frac{1}{2}}\alpha_{i,j-1} + j^{-\frac{1}{2}}\int_0^\infty \phi_{2i}(y)\phi_{2j-1}(y)\,dy$$

$$= \{(2j-1)/(2j)\}^{\frac{1}{2}}\alpha_{i,j-1} + (-1)^{i+j+1}\pi^{-\frac{1}{2}} d_i d_j\,/(2j-1-2i)$$

$$= \pi^{-\frac{1}{2}} \sum_{k=i+1}^{j} (-1)^{i+k+1} \left\{\prod_{\ell=k+1}^{j} \{(2\ell-1)/(2\ell)\}^{\frac{1}{2}}\right\} d_i d_k/(2k-1-2i)$$

$$+ \left\{\prod_{\ell=i+1}^{j} \{(2\ell-1)/(2\ell)\}^{\frac{1}{2}}\right\}\alpha_{ii}$$

where by definition $\prod_{\ell=j+1}^{j} = 1$ and $\sum_{\ell=j+1}^{j} = 0$.

However, since $d_k \prod_{\ell=k+1}^{j} \{(2\ell-1)/(2\ell)\}^{\frac{1}{2}} = d_j$ and $\lambda_{ii} = 0$, we have, if $j > i$,

$$\alpha_{ij} = \pi^{-\frac{1}{2}} d_i d_j \sum_{k=i+1}^{j} (-1)^{i+k+1}/(2k-1-2i) + \tfrac{1}{4}\pi^{\frac{1}{2}} d_i d_j$$

and

$$\lambda_{ij} = 2\pi^{-\frac{1}{2}} d_i d_j \sum_{k=1}^{j-i} (-1)^{k+1}/(2k-1)\,.$$

From the definition of μ_{ij} we have

$$\mu_{ij} = -\tfrac{1}{2}\int_0^\infty \phi'_{2i}(y)\phi_{2j}(y)\,dy - \tfrac{1}{4}\phi_{2i}(0)\phi_{2j}(0).$$

Using (iv) and (v) we obtain

$$\int_0^\infty \phi'_{2i}(y)\phi_{2j}(y)\,dy = 2^{-\frac{1}{2}}\int_0^\infty \{(2i)^{\frac{1}{2}}\phi_{2i-1}(y) - (2i+1)^{\frac{1}{2}}\phi_{2i+1}(y)\}\,\phi_{2j}(y)\,dy$$

$$= (-1)^{i+j+1}\,\pi^{-\frac{1}{2}}\,d_j\{id_i/(2i-1-2j) + (i+\tfrac{1}{2})d_i/(2i+1-2j)\}.$$

Hence

$$\mu_{ij} = (-1)^{i+j}\,\tfrac{1}{2}\,\pi^{-\frac{1}{2}}\,d_i d_j\,(i-j)(2i+2j+1)/\{(2i-2j-1)(2i-2j+1)\}.$$

Finally, from (i), (ii) and (iii),

$$g_{ij} = \int_0^\infty \phi_{2i}(y)\phi_{2j}(y)\,dy - \tfrac{1}{2}\phi_{2j}(0)\int_0^\infty \phi_{2i}(y)\,dy$$

$$= \tfrac{1}{2}\delta_{ij} - (-1)^j\,2^{-3/2}\,d_i d_j\,.$$

In the case of $n = 4$ for example, it is easily seen that

$$p_4 = |g_{oo}g_{11} - \mu_{o1}\lambda_{o1} + g_{o1}^2|$$

$$= |(\tfrac{1}{2} - 2^{-3/2})(\tfrac{1}{2} + 2^{-5/2}) - (2\pi)^{-1} + 1/16|$$

$$= \tfrac{1}{4} - 2^{-7/2} - (2\pi)^{-1}.$$

PROOF OF LEMMA 6.5

1. Let W be an $n\times n$ symmetric matrix. Then

$$K'_n H_n A\otimes A K'_n \operatorname{vech} W = K'_n H_n (A\otimes A)\operatorname{vec} W = K'_n H_n \operatorname{vec}(A W A')$$

$$= K'_n \operatorname{vech}(A W A') = \operatorname{vec}(A W A')$$

$$= (A\otimes A)\operatorname{vec} W$$

$$= (A\otimes A)\,K'_n \operatorname{vech} W\,.$$

Thus, since $\{\operatorname{vech} W : W \text{ is } n\times n \text{ symmetric}\} = \mathbb{R}^{n(n+1)/2}$, it follows that $K'_n H_n (A\otimes A) K'_n = (A\otimes A) K'_n$.

2. Let W be an $n \times n$ symmetric matrix and Z the $n \times n$ symmetric matrix given by $Z = AWA'$. Then, from the proof of part 1 of the lemma,

$$H_n(A \otimes A) K_n' \operatorname{vech} W = \operatorname{vech}(AWA') = \operatorname{vech} Z .$$

But $\operatorname{vec} Z = A \otimes A \operatorname{vec} W$ and so $\operatorname{vec} W = A^{-1} \otimes A^{-1} \operatorname{vec} Z$. Thus

$$\operatorname{vech} W = H_n(A^{-1} \otimes A^{-1}) \operatorname{vec} Z = H_n(A^{-1} \otimes A^{-1}) K_n' \operatorname{vech} Z .$$

Hence the inverse of $H_n(A \otimes A) K_n'$ exists and is given by $H_n(A^{-1} \otimes A^{-1}) K_n'$. #

CHAPTER 7

THE ESTIMATION OF MULTIVARIATE MODELS

7.1 Preliminary

From the point of view of estimation, to date we have only considered scalar RCA models. In this chapter we shall give a brief theoretical discussion of the estimation of multivariate RCA models. While the least squares estimation procedure of chapter 3 will be seen to extend readily to the multivariate situation, the extension of the maximum likelihood procedure is not as straightforward.

For reasons of efficiency we shall, in this chapter, make slight notational changes, writing X_t, ε_t, Y_t, B_t and η_t for $X(t)$, $\varepsilon(t)$, $Y(t)$, $B(t)$ and $\eta(t)$ respectively. As well as the conditions (i)-(iv) assumed in chapter 2, and the assumptions (v) and (vi) of chapter 3, we shall need to assume the following condition.

(vii) Letting $Z_t = K_m[(Y_{t-1} \otimes I_p) \otimes (Y_{t-1} \otimes I_p)]H_p'$, where $m = np^2$ and K_m and H_p are defined in theorem A.1.3, then $E[(Z_t-E(Z_t))(Z_t-E(Z_t))']$ is positive definite if $\{X_t\}$ has finite fourth moments.

It should be noted that if a unique F_t measurable second order stationary solution $\{X_t\}$ exists to (1.1.1), where F_t is the σ-field generated by the set $\{(\varepsilon_t,B_t),(\varepsilon_{t-1},B_{t-1}),\ldots\}$, then X_t is also strictly stationary and ergodic by theorem 2.7.

7.2 The Least Squares Estimation Procedure

In this section we shall obtain a least squares procedure for the estimation of the unknown parameters of a multivariate RCA model.

Letting $\beta = (\beta_n,\ldots,\beta_1)$, then (1.1.1) may be rewritten as

$$(7.2.1)\qquad X_t = \sum_{i=1}^{n} \beta_i X_{t-i} + u_t = \beta Y_{t-1} + u_t = (Y'_{t-1} \otimes I_p)\mathrm{vec}(\beta) + u_t$$

where

$$(7.2.2)\qquad u_t = \sum_{i=1}^{n} B_i(t) X_{t-i} + \varepsilon_t = B_t Y_{t-1} + u_t = (Y'_{t-1} \otimes I_p)\mathrm{vec}(B_t) + \varepsilon_t .$$

Let $\Sigma = E\{\mathrm{vec}(B_t)\mathrm{vec}'(B_t)\}$, so that $\mathrm{vec}\,\Sigma = E\{\mathrm{vec}(B_t) \otimes \mathrm{vec}(B_t)\}$. Since Σ and G are symmetric, the unknown parameters to be estimated are the $(m \times 1)$ vector of coefficients $\mathrm{vec}\,\beta$, and the vectors of covariances $\mathrm{vech}\,\Sigma$ and $\mathrm{vech}\,G$ of dimensions $r = m(m+1)/2$ and $v = p(p+1)/2$ respectively.

From (7.2.2), with F_t defined in §7.1,

$$\begin{aligned} E(u_t u'_t | F_{t-1}) &= (Y'_{t-1} \otimes I_p) E[\mathrm{vec}(B_t)\mathrm{vec}'(B_t)](Y_{t-1} \otimes I_p) + G \\ &= (Y'_{t-1} \otimes I_p)\Sigma(Y_{t-1} \otimes I_p) + G. \end{aligned}$$

But

$$\begin{aligned} \mathrm{vec}[(Y'_{t-1} \otimes I_p)\Sigma(Y_{t-1} \otimes I_p)] &= [(Y'_{t-1} \otimes I_p) \otimes (Y'_{t-1} \otimes I_p)]\mathrm{vec}\,\Sigma \\ &= [\{\mathrm{vec}'(Y_{t-1}Y'_{t-1})\} \otimes I_s]\mathrm{vec}\,\Sigma \end{aligned}$$

where $s = p^2$, and

$$(7.2.3)\qquad \begin{aligned} \mathrm{vech}\,E(u_t u'_t | F_{t-1}) &= H_p[(Y_{t-1} \otimes I_p) \otimes (Y_{t-1} \otimes I_p)]'K'_m \,\mathrm{vech}\,\Sigma + \mathrm{vech}\,G \\ &= Z'_t\,\sigma + g , \quad \text{say} \end{aligned}$$

where $\sigma = \mathrm{vech}\,\Sigma$ and $g = \mathrm{vech}\,G$.

Given the sample of size $(N+n)$, say $X_{-n+1},\dots,X_N$, the estimates of the unknown parameters are obtained in the following manner. From (7.2.1) we may use least squares to obtain the estimate $\hat{\beta}$ of β by minimizing $\sum_{t=1}^{N} u'_t u_t$ with respect to $\mathrm{vec}(\beta)$ i.e. by minimizing

$$S = \sum_{t=1}^{N} [X_t - (Y'_{t-1} \otimes I_p)\text{vec}\,\beta]' [X_t - (Y'_{t-1} \otimes I_p)\text{vec}\,\beta] .$$

But

$$\frac{\partial S}{\partial\, \text{vec}\,\beta} = \sum_{t=1}^{N} [-2(Y_{t-1} \otimes I_p)X_t + 2(Y_{t-1} \otimes I_p)(Y'_{t-1} \otimes I_p)\text{vec}\,\beta]$$

so that

$$\text{vec}\,\hat{\beta} = [\sum_{t=1}^{N} (Y_{t-1} \otimes I_p)(Y'_{t-1} \otimes I_p)]^{-1}[\sum_{t=1}^{N} (Y_{t-1} \otimes I_p)X_t]$$

$$= [(\sum_{t=1}^{N} Y_{t-1}Y'_{t-1})^{-1} \otimes I_p][\sum_{t=1}^{N} (Y_{t-1} \otimes I_p)X_t] .$$

Using theorem A.1.1 it follows that

(7.2.4) $$\text{vec}\,\hat{\beta} = [(\sum_{t=1}^{N} Y_{t-1}Y'_{t-1})^{-1} \otimes I_p] \sum_{t=1}^{N} \text{vec}(X_t Y'_{t-1})$$

$$= \sum_{t=1}^{N} \text{vec}[X_t Y'_{t-1}(\sum_{t=1}^{N} Y_{t-1}Y'_{t-1})^{-1}]$$

and so

$$\hat{\beta} = (\sum_{t=1}^{N} X_t Y'_{t-1})(\sum_{t=1}^{N} Y_{t-1}Y'_{t-1})^{-1}.$$

Having estimated β, from (7.2.1) we may estimate u_t by

$$\hat{u}_t = X_t - (Y'_{t-1} \otimes I_p)\text{vec}\,\hat{\beta} .$$

From (7.2.3) estimates of σ and g may be found by regressing $\text{vech}(\hat{u}_t\hat{u}'_t)$ on Z'_t and I_v, where $v = p(p+1)/2$. Thus letting

(7.2.5) $$\hat{e}_t = \text{vech}(\hat{u}_t\hat{u}'_t) - g - Z'_t\,\sigma$$

the estimates of σ and g are obtained by minimizing $\sum_{t=1}^{N} \hat{e}'_t\hat{e}_t$, and are given by

$$(7.2.6) \qquad \hat{\sigma} = \text{vech}\,\hat{\Sigma} = [\sum_{t=1}^{N} (Z_t-\bar{Z})(Z_t-\bar{Z})']^{-1}[\sum_{t=1}^{N} (Z_t-\bar{Z})\,\text{vech}\,\hat{u}_t\hat{u}_t']$$

and hence, from (7.2.3), the $v \times 1$ vector g is estimated by

$$(7.2.7) \qquad \hat{g} = \text{vech}\,\hat{G} = N^{-1} \sum_{t=1}^{N} \text{vech}(\hat{u}_t\hat{u}_t') - \bar{Z}'\hat{\sigma}$$

where $\bar{Z} = N^{-1} \sum_{t=1}^{N} Z_t$.

7.3 The Asymptotic Properties of the Estimates

In order to derive the strong consistency and asymptotic normality of the least squares estimates (7.2.4), (7.2.6) and (7.2.7) it is convenient to consider first the asymptotic theory for $\text{vec}\,\hat{\beta}$. Having done this we may then consider the properties of the elements of the estimated covariance matrices $\hat{\Sigma}$ and $\hat{G}$.

THEOREM 7.1 Consider the process $\{X_t\}$ which satisfies (7.1.1) subject to the assumptions (i)-(vii). If $E(X_t^2(i)) < \infty$, for $i = 1,\ldots,p$, where $X_t = (X_t(1),\ldots,X_t(p))'$, then $\hat{\beta}$ defined by (7.2.4) is a strongly consistent estimate of β. Furthermore, if $E(X_t^4(i)) < \infty$ then $N^{1/2}\text{vec}(\hat{\beta}-\beta)$ has a distribution which converges to that of a normally distributed random vector with zero mean and covariance matrix

$$(7.3.1) \qquad V^{-1} \otimes G + W^{-1}E[W_t\,\Sigma\,W_t]W^{-1}$$

where

$$V = E(Y_{t-1}Y_{t-1}'), \quad W = V \otimes I_p \quad \text{and} \quad W_t = (Y_{t-1}Y_{t-1}') \otimes I_p .$$

Proof. From (7.2.4) it follows that

$$(7.3.2) \qquad \text{vec}(\hat{\beta}-\beta) = \left[(\sum_{t=1}^{N} Y_{t-1}Y_{t-1}')^{-1} \otimes I_p\right]\left[\sum_{t=1}^{N} (Y_{t-1} \otimes I_p)u_t\right].$$

By the ergodic theorem

$$N^{-1} \sum_{t=1}^{N} (Y_{t-1}Y'_{t-1}) \otimes I_p \overset{a.s.}{\rightarrow} [E(Y_{t-1}Y'_{t-1})] \otimes I_p = V \otimes I_p$$

so that $\text{vec}(\hat{\beta} - \beta)$ has the same asymptotic properties as

$$(V^{-1} \otimes I_p)N^{-1} \sum_{t=1}^{N} (Y_{t-1} \otimes I_p)u_t .$$

It is easily seen that

$$\begin{aligned} E[(Y_{t-1} \otimes I_p)u_t | F_{t-1}] &= (Y_{t-1} \otimes I_p)E[\{Y'_{t-1} \otimes I_p)\text{vec}(B_t) + \varepsilon_t\} | F_{t-1}] \\ &= 0 , \end{aligned}$$

and since Y_{t-1} and u_t are both strictly stationary and ergodic, so is $(Y_{t-1} \otimes I_p)u_t$. Hence again using the ergodic theorem we have, since $E(X_t^2(i)) < \infty$, $i = 1,\dots,p$,

$$N^{-1} \sum_{t=1}^{N} (Y_{t-1} \otimes I_p)u_t \overset{a.s.}{\rightarrow} 0$$

and so, from (7.3.2), it follows that $\text{vec}(\hat{\beta} - \beta) \overset{a.s.}{\rightarrow} 0$ i.e. the estimate $\text{vec}\,\hat{\beta}$ of $\text{vec}\,\beta$ defined by (7.2.4) is strongly consistent.

If α is an $m \times 1$ constant vector then by theorem A.1.4,

$$N^{-\frac{1}{2}} \sum_{t=1}^{N} \alpha'(Y_{t-1} \otimes I_p)u_t$$

has a distribution which converges to the normal distribution with zero mean and variance

$$E[\{\alpha'(Y_{t-1}\otimes I_p)u_t\}^2] = E[\alpha'(Y_{t-1}\otimes I_p)E(u_tu_t'|F_{t-1})(Y_{t-1}'\otimes I_p)\alpha]$$

$$= E[\alpha'(Y_{t-1}\otimes I_p)\{(Y_{t-1}'\otimes I_p)\Sigma(Y_{t-1}\otimes I_p)+G\}(Y_{t-1}'\otimes I_p)\alpha]$$

$$= \alpha' E[W_t\Sigma W_t + (Y_{t-1}\otimes I_p)G(Y_{t-1}'\otimes I_p)]\alpha ,$$

$$= \alpha' E[W_t\Sigma W_t)\alpha + \alpha'(\{E(Y_{t-1}Y_{t-1}')\}\otimes G)\alpha = \alpha' E[W_t\Sigma W_t]\alpha + \alpha'(V\otimes G)\alpha$$

provided $E(X_t^4(i)) < \infty$, for $i = 1,\dots,p$, where $X_t = (X_t(1),\dots,X_t)p))'$.

Thus $N^{\frac{1}{2}}\text{vec}(\hat\beta - \beta)$ converges in distribution to the normal distribution with zero mean and covariance matrix given by (7.3.1), as required. #

It will prove convenient to introduce quantities $\tilde\sigma$ and $\tilde g$ defined in the same way as $\hat\sigma$ and $\hat g$ respectively except that $\hat u_t$ has been replaced by u_t in the respective formulae, which will be used in proving the following lemma needed to obtain the asymptotic properties of $\hat\sigma$ and $\hat g$.

LEMMA 7.1 <u>If</u> $E(X_t^4(i)) < \infty$, $i = 1,\dots,p$ <u>where</u> $X_t' = (X_t(1),\dots,X_t(p))$, <u>then</u> $(\tilde\sigma - \hat\sigma)$ <u>and</u> $(\tilde g - \hat g)$ <u>converge almost surely to zero, while</u> $N^{\frac{1}{2}}(\tilde\sigma - \hat\sigma)$ <u>and</u> $N^{\frac{1}{2}}(\tilde g - \hat g)$ <u>converge in probability to zero</u>.

<u>Proof.</u> See appendix 7.1.

For convenience we define the vector $\lambda = (\text{vec}'(\beta),\sigma',g')'$ with $\hat\lambda$ and $\tilde\lambda$ defined in an obvious way and $\tilde\beta = \hat\beta$. The asymptotic properties of $\hat\lambda$ are obtained in the following theorem.

THEOREM 7.2 <u>Under the same conditions as for theorem 7.1,</u> $(\hat\lambda - \lambda)$ <u>converges almost surely to zero if</u> $E(X_t^4(i)) < \infty$, <u>while if</u> $E(X_t^8(i)) < \infty$, $N^{\frac{1}{2}}(\hat\lambda - \lambda)$ <u>has a distribution which converges to that of a normally distributed random vector with mean zero and covariance matrix</u> S <u>defined by (7.3.5)</u>.

<u>Proof.</u> In view of lemma 7.1 all that needs to be done is to consider the corresponding results for $\tilde\lambda = (\text{vec}'\tilde\beta,\tilde\sigma',\tilde g')'$. Now from (7.2.5) and (7.2.6) we have

$$(7.3.3) \qquad \tilde{\sigma} - \sigma = [\sum_{s=1}^{N} (Z_s-\bar{Z})(Z_s-\bar{Z})']^{-1} \sum_{t=1}^{N} (Z_t-\bar{Z})e_t$$

where

$$e_t = \text{vech}(u_t u_t') - g - Z_t'\sigma ,$$

while if

$$\nu_t = I_v - \bar{Z}'[N^{-1} \sum_{s=1}^{N} (Z_s-\bar{Z})(Z_s-\bar{Z})']^{-1}(Z_t-\bar{Z})$$

then from (7.2.5) and (7.2.7)

$$(7.3.4) \qquad \tilde{g} - g = N^{-1} \sum_{t=1}^{N} \text{vech}(u_t u_t') - \bar{Z}'\tilde{\sigma} - g = N^{-1} \sum_{t=1}^{N} \nu_t e_t .$$

We shall need to replace the terms $\bar{Z}$ and $N^{-1} \sum_{s=1}^{N} (Z_s-\bar{Z})(Z_s-\bar{Z})'$ which occur in (7.3.3) and (7.3.4) by their limits.

Now

$$N^{-1} \sum_{t=1}^{N} (Z_t-\bar{Z})e_t = N^{-1} \sum_{t=1}^{N} (Z_t-E(Z_t))e_t + N^{-1} \sum_{t=1}^{N} (E(Z_t)-\bar{Z})e_t .$$

By the ergodic theorem $\bar{Z} \overset{a.s.}{\rightarrow} E(Z_t)$ if $E(X_t^2(i)) < \infty$ and

$$N^{-1} \sum_{s=1}^{N} (Z_s-\bar{Z})(Z_s-\bar{Z})' \overset{a.s.}{\rightarrow} E\{(Z_s-E(Z_s))(Z_s-E(Z_s))'\} \text{ if } E(X_t^4(i)) < \infty .$$

Furthermore, $N^{-1} \sum_{t=1}^{N} (E(Z_t)-\bar{Z})e_t = (E(Z_t)-\bar{Z})\bar{e}$, where $\bar{e} = N^{-1} \sum_{t=1}^{N} e_t$. But

$$E(e_t|F_{t-1}) = E[(\text{vech}(u_t u_t') - Z_t'\sigma - g)|F_{t-1}] = 0$$

so that $\bar{e} \overset{a.s.}{\rightarrow} 0$ if $E(X_t^2(i)) < \infty$. Thus by theorem A.1.4 if $E(X_t^4(i)) < \infty$ it follows that $N^{\frac{1}{2}}\bar{e}$ has a distribution which converges to that of a normally distributed random vector with zero mean, and so

$$(E(Z_t)-\overline{Z})\overline{e} \overset{a.s.}{\rightarrow} 0 \quad \text{and} \quad N^{\frac{1}{2}}\{E(Z_t)-\overline{Z}\}\overline{e} \overset{p}{\rightarrow} 0 .$$

If we now let $\lambda^* = (\text{vec}'\ \beta^*,\ \sigma^{*\prime},\ g^{*\prime})'$ where

$$\text{vec}(\beta^*-\beta) = W^{-1}N^{-1}\sum_{t=1}^{N}(Y_{t-1}\otimes I_p)u_t\ , \quad \sigma^*-\sigma = R^{-1}N^{-1}\sum_{t=1}^{N}(Z_t-E(Z_t))e_t$$

and

$$g^*-g = N^{-1}\sum_{t=1}^{N}\nu_t^* e_t$$

with $R = E[(Z_t-E(Z_t))(Z_t-E(Z_t))']$ and $\nu_t^* = I_v - E(Z_t')R^{-1}(Z_t-E(Z_t))$ then, from above, it immediately follows that $(\lambda^*-\tilde{\lambda}) \overset{a.s.}{\rightarrow} 0$ and $N^{\frac{1}{2}}(\lambda^*-\tilde{\lambda}) \overset{p}{\rightarrow} 0$. Consequently $(\lambda^*-\hat{\lambda}) \overset{a.s.}{\rightarrow} 0$ and $N^{\frac{1}{2}}(\lambda^*-\hat{\lambda}) \overset{p}{\rightarrow} 0$ so the asymptotic results for $\hat{\lambda}$ may be derived from those of λ^*.

If the vector of constants $a = (a_1',a_2',a_3')'$ is such that a_1 is $m\times 1$ a_2 is $r\times 1$ and a_3 is $v\times 1$, then

$$a'(\lambda^*-\lambda) = N^{-1}\sum_{t=1}^{N} M_t(a)\ ,$$

where

$$M_t(a) = a_1'W^{-1}(Y_{t-1}\otimes I_p)u_t + a_2'R^{-1}(Z_t-E(Z_t))e_t$$
$$+ a_3'[I_v-E(Z_t')R^{-1}(Z_t-E(Z_t))]e_t\ .$$

But $E(M_t(a)|F_{t-1}) = 0$, $M_t(a)$ is strictly stationary and ergodic and $E(M_t^2(a))$ exists when $E(X_t^8(i)) < \infty$. Thus $a'(\lambda^*-\lambda) \overset{a.s.}{\rightarrow} 0$ and so $(\lambda^*-\lambda) \overset{a.s.}{\rightarrow} 0$, as required.

If $E(X_t^8(i)) < \infty$, $N^{\frac{1}{2}}a'(\lambda^*-\lambda)$ has a distribution which converges to that of a normally distributed random variable with mean zero and variance $E(M_t^2(a))$. We can however write $E(M_t^2(a)) = a'Sa$, say. Thus $N^{\frac{1}{2}}(\lambda^*-\lambda)$ and hence $N^{\frac{1}{2}}(\lambda^*-\lambda)$ has a distribution which converges to that of a normally distributed random vector with mean zero and covariance matrix S, where

(7.3.5) $$S = \{S_{ij}\}, \qquad i,j = 1,2,3.$$

The block matrix S_{ij} is obtained by evaluating the corresponding component of the form $a_i'S_{ij}a_j$ in the expansion of $E(M_t^2(a))$. Thus S_{11} is given by (7.3.1) while

$$S_{12} = S_{21}' = W^{-1}E[(Y_{t-1} \otimes I_p)E\{u_tu_t'(u_t' \otimes I_p)|F_{t-1}\}H_p'(Z_t-E(Z_t))']R^{-1}$$

$$S_{13} = S_{31}' = W^{-1}E[(Y_{t-1} \otimes I_p)E\{u_tu_t'(u_t' \otimes I_p)|F_{t-1}\}H_p'\{I_v - (Z_t-E(Z_t))'R^{-1}E(Z_t)\}]$$

$$S_{22} = R^{-1}E[(Z_t-E(Z_t))E(e_te_t'|F_{t-1})(Z_t-E(Z_t))']R^{-1}$$

$$S_{23} = S_{32}' = R^{-1}E[(Z_t-E(Z_t))E(e_te_t'|F_{t-1})\{I_v - (Z_t-E(Z_t))'R^{-1}E(Z_t)\}]$$

and

$$S_{33} = E\{[I_v-E(Z_t')R^{-1}(Z_t-E(Z_t))]E(e_te_t'|F_{t-1})[I_v-(Z_t-E(Z_t))'R^{-1}E(Z_t)]\}$$

where

$$E(e_te_t'|F_{t-1})$$

$$= E[\{H_p(u_t \otimes I_p)u_tu_t'(u_t' \otimes I_p)H_p' - (\sigma+Z_t'\omega)(\sigma+Z_t'\omega)'\}|F_{t-1}] .$$

$$= E[\{H_p(u_tu_t') \otimes (u_tu_t')H_p' - (\sigma+Z_t'\omega)(\sigma+Z_t'\omega)'\}|F_{t-1}] .$$

7.4 Maximum Likelihood Estimation

As in the scalar case, given a sample $\{X_1,\ldots,X_N\}$ from a multivariate stationary ergodic time series $\{X_t\}$ which satisfies (1.1.1) and conditions (i)-(vii), we shall derive the likelihood function, conditional on the preperiod values $\{X_0,\ldots,X_{1-n}\}$, as though we were assuming the joint normality of $\{\varepsilon_t\}$ and $\{B_t\}$. Now

$$E(X_t | Y_{t-1}) = \beta Y_{t-1}$$

and, as was shown in §7.2, the covariance matrix of X_t given Y_{t-1} is given by

$$E(u_t u_t' | Y_{t-1}) = G + (Y_{t-1}' \otimes I_p)\Sigma(Y_{t-1} \otimes I_p) = \Omega_t \text{ , say.}$$

Hence letting $f_N(X_1, \ldots, X_N | X_o, \ldots, X_{1-n})$ be the conditional density of $X_1, \ldots, X_N$ given $X_o, \ldots, X_{1-n}$, it follows that

$$(7.4.1) \qquad f_N(X_1, \ldots, X_N | X_o, \ldots, X_{1-n}) = \prod_{t=1}^{N} f(X_t | X_{t-1}, \ldots, X_{t-n})$$

$$= (2\pi)^{-\frac{Np}{2}} \prod_{t=1}^{N} [|\Omega_t|^{-\frac{1}{2}} \exp\{-\tfrac{1}{2}(X_t - \beta Y_{t-1})' \Omega_t^{-1} (X_t - \beta Y_{t-1})\}] .$$

Now

$$\sum_{t=1}^{N} (X_t - \beta Y_{t-1})' \Omega_t^{-1} (X_t - \beta Y_{t-1})$$

$$= \sum_{t=1}^{N} \{X_t - (Y'_{t-1} \otimes I_p) \operatorname{vec} \beta\}' \Omega_t^{-1} \{X_t - (Y_{t-1}' \otimes I_p) \operatorname{vec} \beta\}$$

$$= \sum_{t=1}^{N} \{X_t' \Omega_t^{-1} X_t - 2X_t' \Omega_t^{-1} (Y_{t-1}' \otimes I_p) \operatorname{vec} \beta$$

$$+ \operatorname{vec}' \beta (Y_{t-1} \otimes I_p) \Omega_t^{-1} (Y_{t-1}' \otimes I_p) \operatorname{vec} \beta\} .$$

Thus the likelihood function, conditional on $X_o, \ldots, X_{1-n}$, is given by

$$L_N(\beta, G, \Sigma) = f_N(X_1, \ldots, X_N | X_o, \ldots, X_{1-n})$$

$$= (2\pi)^{-\frac{Np}{2}} \prod_{t=1}^{N} [|W_t|^{-\frac{1}{2}} \exp\{-\tfrac{1}{2}(X_t' \Omega_t^{-1} X_t - 2X_t' \Omega_t^{-1} (Y_{t-1}' \otimes I_p) \operatorname{vec} \beta$$

$$+ \operatorname{vec}' \beta (Y_{t-1} \otimes I_p) \Omega_t^{-1} (Y_{t-1}' \otimes I_p) \operatorname{vec} \beta\}] .$$

It is more convenient however to minimize the function

$$(7.4.2) \qquad \tilde{\ell}_N(\beta, G, \Sigma) = -\frac{2}{N} \log L_N(\beta, G, \Sigma) - p \log 2\pi$$

$$= N^{-1}\sum_{t=1}^{N}\log|\Omega_t| + N^{-1}\sum_{t=1}^{N}\{X_t'\Omega_t^{-1}X_t - 2X_t'\Omega_t^{-1}(Y_{t-1}'\otimes I_p)\mathrm{vec}\,\beta$$
$$+ \mathrm{vec}'\beta(Y_{t-1}\otimes I_p)\Omega_t^{-1}(Y_t'\otimes I_p)\mathrm{vec}\,\beta\}.$$

As with estimation for scalar RCA's, the minimization of $\tilde{\ell}_N(\beta,G,\Sigma)$ may be carried out in stages. The function $\tilde{\ell}_N$ may first be minimized with respect to β, the minimizing value being a function of the data and the parameters G and Σ. This minimizer may then be used to "concentrate out" the dependence of $\tilde{\ell}_N$ on the parameter β, giving a function to be minimized which depends only on the data and the parameters G and Σ.

From (7.4.2)

$$\frac{\partial\tilde{\ell}_N}{\partial\mathrm{vec}\,\beta} = -2\sum_{t=1}^{N}(Y_{t-1}\otimes I_p)\Omega_t^{-1}X_t + 2\sum_{t=1}^{N}(Y_{t-1}\otimes I_p)\Omega_t^{-1}(Y_{t-1}'\otimes I_p)\mathrm{vec}\,\beta$$

so that $\tilde{\ell}_N$ is minimized with respect to β when

$$\text{(7.4.3)}\quad \mathrm{vec}\,\beta = \mathrm{vec}\,\beta(G,\Sigma) = [\sum_{t=1}^{N}(Y_{t-1}\otimes I_p)\Omega_t^{-1}(Y_{t-1}'\otimes I_p)]^{-1}[\sum_{t=1}^{N}(Y_{t-1}\otimes I_p)\Omega_t^{-1}X_t]$$
$$= R^{-1}(G,\Sigma)\,S(G,\Sigma)\,,$$

say, and $\Omega_t = \Omega_t(G,\Sigma)$. We now replace $\mathrm{vec}\,\beta$ by $\mathrm{vec}\,\beta(G,\Sigma)$ in $\tilde{\ell}_N$ and obtain the estimates $\hat{G}$, $\hat{\Sigma}$ of G and Σ by minimizing the resulting expression. Thus to obtain $\hat{G}$ and $\hat{\Sigma}$ we must minimize

$$\tilde{\ell}_N(G,\Sigma) = N^{-1}\sum_{t=1}^{N}\log|\Omega_t| + N^{-1}\sum_{t=1}^{N}\{X_t'\Omega_t^{-1}X_t$$
$$- 2X_t'\Omega_t^{-1}(Y_{t-1}'\otimes I_p)\mathrm{vec}\,\beta + \mathrm{vec}'\beta(Y_{t-1}\otimes I_p)\Omega_t^{-1}(Y_{t-1}'\otimes I_p)\mathrm{vec}\,\beta\}$$
$$= N^{-1}\sum_{t=1}^{N}\log|\Omega_t| + N^{-1}\sum_{t=1}^{N}X_t'\Omega_t^{-1}X_t - S'R^{-1}S\,,$$

where $\beta = \beta(G,\Sigma)$, given by (7.4.3).

From the form of Ω_t, R and S, $\tilde{\ell}_N(G,\Sigma)$ is non-linear in the parameters G and Σ to be estimated. Consequently $\text{vech}\,\hat{G}$ and $\text{vech}\,\hat{\Sigma}$, and hence $\hat{G}$ and $\hat{\Sigma}$, the maximum likelihood estimates of G and Σ respectively, will be obtained by applying a non-linear optimization procedure to $\tilde{\ell}_N(G,\Sigma)$. Having obtained $\hat{G}$ and $\hat{\Sigma}$, from (7.4.3) we may then obtain the maximum likelihood estimate $\text{vec}\,\hat{\beta}$ of $\text{vec}\,\beta$ from $\hat{\beta} = \beta(\hat{G},\hat{\Sigma})$.

It would seem reasonable to believe that the maximum likelihood estimates of the parameters for the multivariate model under conditions similar to those for the scalar model, will be strongly consistent and satisfy a central limit theorem. The proofs however would be very complicated.

7.5 Conclusion

In this chapter we have demonstrated how the least squares and maximum likelihood estimation procedures developed in chapters 3 and 4, in the case of scalar RCA models, generalize to include multivariate RCA models. When considering practical multivariate estimation problems one should however be careful, since the number of parameters to be estimated increases dramatically as the number of lags in the RCA increases.

APPENDIX 7.1

PROOF OF LEMMA 7.1 Using (7.2.6) and (7.2.7) we have

$$(\text{A.7.1}) \qquad \hat{\sigma}-\tilde{\sigma} = \left[\sum_{t=1}^{N} (Z_t-\overline{Z})(Z_t-\overline{Z})'\right]^{-1} \left[\sum_{t=1}^{N} (Z_t-\overline{Z})\,\text{vech}(\hat{u}_t\hat{u}_t'-u_tu_t')\right]$$

and

$$(\text{A.7.2}) \qquad \hat{g}-\tilde{g} = N^{-1}\sum_{t=1}^{N} \text{vech}(\hat{u}_t\hat{u}_t'-u_tu_t') - \overline{Z}'(\hat{\sigma}-\tilde{\sigma}) .$$

By the ergodic theorem, if $E(X_t^2(i)) < \infty$, $\overline{Z} \overset{a.s.}{\to} E(Z_t)$, while if $E(X_t^4(i)) < \infty$, $N^{-1}\sum_{t=1}^{N} Z_tZ_t' \overset{a.s.}{\to} E(Z_tZ_t')$ so that

$$N^{-1}\sum_{t=1}^{N} (Z_t-\overline{Z})(Z_t-\overline{Z})' \overset{a.s.}{\to} E[(Z_t-E(Z_t))(Z_t-E(Z_t))'] .$$

Now

$$(\text{A.7.3}) \qquad (\hat{u}_t\hat{u}_t'-u_tu_t') = (\hat{u}_t+u_t)(\hat{u}_t-u_t)' + \hat{u}_tu_t' - u_t\hat{u}_t' .$$

But

$$\begin{aligned} \hat{u}_tu_t' &= \{X_t-(Y_{t-1}'\otimes I_p)\text{vec}\,\hat{\beta}\}\,u_t' \\ &= [X_t-(Y_{t-1}'\otimes I_p)\text{vec}\,\beta - (Y_{t-1}'\otimes I_p)\text{vec}(\hat{\beta}-\beta)]\,u_t' \\ &= u_tu_t' - (Y_{t-1}'\otimes I_p)\text{vec}(\hat{\beta}-\beta)\,u_t' . \end{aligned}$$

Similarly $u_t\hat{u}_t' = u_tu_t' - u_t[\text{vec}(\hat{\beta}-\beta)]'(Y_{t-1}\otimes I_p)$ so that

$$(\text{A.7.4}) \qquad \text{vech}(\hat{u}_tu_t'-u_t\hat{u}_t') = H_p[(Y_{t-1}'\otimes I_p)\otimes u_t - u_t\otimes(Y_{t-1}'\otimes I_p)]\text{vec}(\hat{\beta}-\beta) .$$

Furthermore

$$\begin{aligned} (\hat{u}_t+u_t) &= 2X_t-(Y_{t-1}'\otimes I_p)\text{vec}(\hat{\beta}+\beta) \\ &= 2[X_t-(Y_{t-1}'\otimes I_p)\text{vec}\,\beta] - (Y_{t-1}'\otimes I_p)\text{vec}(\hat{\beta}-\beta) \end{aligned}$$

$$= 2u_t - (Y'_{t-1} \otimes I_p)\text{vec}(\hat{\beta}-\beta)$$

and

$$(\hat{u}_t - u_t) = -(Y'_{t-1} \otimes I_p)\text{vec}(\hat{\beta}-\beta) .$$

Thus

$$\text{vech}[(\hat{u}_t+u_t)(\hat{u}_t-u_t)] = \text{vech}\{[-(2u_t-(Y'_{t-1} \otimes I_p)\text{vec}(\hat{\beta}-\beta)]\{\text{vec}'(\hat{\beta}-\beta)\}(Y_{t-1} \otimes I_p)\}$$

$$= H_p\text{vec}[(Y'_{t-1} \otimes I_p)\{\text{vec}(\hat{\beta}-\beta)\text{vec}'(\hat{\beta}-\beta)\}(Y_{t-1} \otimes I_p)-2u_t\{\text{vec}'(\hat{\beta}-\beta)\}(Y_{t-1} \otimes I_p)]$$

$$= H_p[\{(Y'_{t-1} \otimes I_p) \otimes (Y'_{t-1} \otimes I_p)\}\text{vec}\{\text{vec}(\hat{\beta}-\beta)\text{vec}'(\hat{\beta}-\beta)\}$$

$$-2\{(Y'_{t-1} \otimes I_p) \otimes u_t\}\text{vec}\{\text{vec}'(\hat{\beta}-\beta)\}] .$$

But

$$\text{vec}\{\text{vec}(\hat{\beta}-\beta)\text{vec}'(\hat{\beta}-\beta)\} = [I_m \otimes \text{vec}(\hat{\beta}-\beta)]\text{vec}\{\text{vec}'(\hat{\beta}-\beta)\}$$

$$= [I_m \otimes \text{vec}(\hat{\beta}-\beta)]\text{vec}(\hat{\beta}-\beta)$$

where, as before, $m = np^2$. Consequently

$$\text{(A.7.5)} \qquad \text{vech}[(\hat{u}_t+u_t)(\hat{u}_t-u_t)']$$

$$= H_p[\{Y'_{t-1} \otimes I_p) \otimes (Y'_{t-1} \otimes I_p)\}\{I_m \otimes \text{vec}(\hat{\beta}-\beta)\}-2\{(Y'_{t-1} \otimes I_p) \otimes u_t\}]\text{vec}(\hat{\beta}-\beta) .$$

It follows from the ergodic theorem that, if $E(X_t^4(i)) < \infty$, $N^{-1} \sum_{t=1}^{N} (Z_t-\bar{Z})H_p[(Y'_{t-1} \otimes I_p) \otimes u_t] \overset{a.s.}{\to} 0$. Furthermore $\text{vec}(\hat{\beta}-\beta) \overset{a.s.}{\to} 0$ and if $E(X_t^4(i)) < \infty$ then from (A.7.1) and (A.7.3)-(A.7.5) it can be seen that $\hat{\sigma} - \tilde{\sigma} \overset{a.s.}{\to} 0$.

From theorem 7.1 it follows that $N^{\frac{1}{2}}(\hat{\beta}-\beta)$ converges in distribution Using the ergodic theorem and theorem A.1.4, from (A.7.1) and (A.7.3)-(A.7.5) we have $N^{\frac{1}{2}}(\hat{\sigma}-\tilde{\sigma}) \overset{p}{\to} 0$.

Using a similar type of argument, from (A.7.2) it follows in a straightforward manner that $(\hat{g}-\tilde{g}) \overset{a.s.}{\rightarrow} 0$ and $N^{\frac{1}{2}}(\hat{g}-\tilde{g}) \overset{p}{\rightarrow} 0$. #

CHAPTER 8

AN APPLICATION

8.1 Introduction

This final chapter will be concerned with the application to real data of the procedures which have been developed for the RCA model. We shall remodel the lynx data, a set of univariate data which has been considered by many authors in the past, and to which several different types of models have been applied. An extensive account of the statistical and historical aspects of the modelling of the lynx data is contained in Campbell and Walker (1977) (where the data are reproduced since the original sources are relatively inaccessible) and the discussion which follows.

Moran (1953) carried out the first rigorous statistical analysis of the lynx data, proposing a second order autoregressive model for the series. Noting that the one-step-ahead predictors for the data were not particularly good, however, he suggested that "perhaps ... the process would be better represented by some kind of non-linear model". Tong (1977) takes the same view and is "reasonably optimistic that ... AR models will provide good first approximations to any future more refined non-linear model". Although Tong has obtained an autoregressive model which includes lags up to and including the eleventh, it would seem relevant to consider, in the light of the model obtained by Moran, the fitting of a second order random coefficient autoregression to the lynx data.

In order to compare several models which have been considered for the data, we re-estimate the models using only the first one hundred observations (there are one hundred and fourteen altogether) and compute various one-step-ahead predictors for the remaining fourteen observations. It is seen from (1.1.1) that, for a random coefficient autoregressive process $\{X(t)\}$, $E(X(t)|F_{t-1}) = \beta'Y(t-1) = \hat{X}(t)$, say. Thus the best predictor, in the least squares sense, of $X(t)$ from $\{X(t-1), X(t-2), \ldots\}$ is a linear predictor. The principle of least squares, however, may not be an appropriate principle in this case, since the non-linear nature of (1.1.1) would usually prevent the process $\{X(t)\}$ from being Gaussian, or even near Gaussian.

In an attempt to exploit this non-linearity, squaring both sides of (1.1.1) it is seen that

$$E(X^2(t)|F_{t-1}) = \{\beta'Y(t-1)\}^2 + \sigma^2 + \gamma'z(t)$$

so that a natural predictor of $X(t)$ from $\{X(t-1), X(t-2), \ldots\}$ is

$$\tilde{X}(t) = \operatorname{sgn}(\beta'Y(t-1))\,[\{\beta'Y(t-1)\}^2 + \sigma^2 + \gamma'z(t)]^{\frac{1}{2}}\,, \tag{8.1.1}$$

where

$$\operatorname{sgn}(x) = \begin{cases} 1, & x \geq 0 \\ -1, & x < 0\,. \end{cases}$$

A similar predictor $X^*(t)$ may be obtained for fixed coefficient autoregressive models by letting $\gamma = 0$ in (8.1.1).

Although the reasoning behind considering a predictor of the form (8.1.1) is somewhat heuristic, it will be seen that, in the case of the lynx data, $\tilde{X}(t)$ is better at predicting than the usual linear predictors. Finally, it should be noted that the expression

$$E\{(X(t) - \beta'Y(t-1))^2|F_{t-1}\} = \sigma^2 + \gamma'z(t)$$

may not provide a non-linear predictor for $X(t)$, since there appears to be no obvious method for producing a sign for $(\sigma^2 + \gamma'z(t))^{\frac{1}{2}}$ which does not involve knowledge of $X(t)$.

8.2 A Non-linear Model for the Lynx Data

The lynx data consists of the annual records of the numbers of Canadian lynx trapped in the Mackenzie River district of North-Western Canada over a period of 114 years between 1821-1934. The data originally appeared in Elton and Nicholson (1942), the authors noticing an apparently regular cycle of approximately ten years in the data. Moran (1953) endeavoured to obtain a statistical model for the data which would explain this regularity. Because the peaks of the data were sharp, while the troughs were relatively smooth, Moran first took the logarithms to base ten of the data, since the models he was to consider induce a more or less symmetric behaviour about the mean. The transformed set has a skewness and kurtosis much closer to those of a Gaussian process, although a noticeable feature of the set is that the ascending sections of the cycle take longer to complete than the descending sections, as may be seen in figure 8.1. Since Moran's priority was to obtain a model explaining the cycle observed in the actual populations, a major assumption he made was that the untransformed data was directly proportional to the true population size.

The lynx has as its main food-source the snowshoe rabbit (or hare), and it is well-known that a predator-prey relationship may induce oscillations in the two relevant populations. The ten-year cycle, however, has been noticed in other species in Canada. Indeed, as noted by Bulmer (1974, 1975), the snowshoe rabbit is the prey of many larger animals throughout Canada, and the ten-year cycle appears to be synchronized within these species over the whole of the country. Bulmer has thus concluded that the lynx has a cycle induced by that of the snowshoe rabbit, but does not affect seriously the size of the snowshoe rabbit population. Finerty (1980) has also considered the relationship between lynx and snowshoe hare. He concludes that "the appropriate comparison of lynx and hare pelts suggests that the phase relationship between lynx and hare is what would be expected from a predator-follows-prey situation".

One hypothesis which Moran considered to explain the cycle was that "the oscillations arise from the population dynamics of the lynx themselves". He thus undertook a dynamic modelling of the data which would explain the cycle, for which the population in any one year would depend on the population of previous years. Hence autoregressive models were fitted to the data, a natural class of models for the data since apparent cycles will arise in an autoregressive process whose characteristic polynomial has complex zeros which are relatively close to the unit circle. Letting $\{X(t);\ t = 1,\ldots,114\}$ denote the transformed data set, Moran obtained, using the Yule-Walker relations, the model

$$(8.2.1)\qquad X(t) - 2.9036 = 1.4101(X(t-1)-2.9036) - .7734(X(t-2)-2.9036) + \varepsilon(t)$$

where the estimated variance of $\varepsilon(t)$ was .0459. The zeros of the characteristic polynomial $(1-1.4101z + .7734z^2)$ are $(.8794)^{-1}\exp\{\pm i2\pi/9.8076\}$, and so one would expect an autoregressive process satisfying (8.2.1) to exhibit cycles with a period of 9.8076. The autocorrelation function of a stationary process satisfying (8.2.1), however, damps out much faster than the autocorrelation function of the data $\{X(t)\}$, suggesting that the model does not provide an adequate fit of the data. In fact Hannan (1960), using a testing procedure due to Quenouille (1947), has shown that the model is far from adequate, although he has also demonstrated that the model is still a more successful alternative than a purely sinusoidal model with a random error term. Bulmer (1974) has consequently fitted a "mixed-spectrum" model, that is, a model containing both an autoregressive term and a sinusoidal component. The model he chooses is of the form

$$(8.2.2)\qquad X(t) - \mu = \alpha \sin[2\pi\omega(t-\phi)] + \beta(X(t-1)-\mu) + \varepsilon(t)$$

where $\omega^{-1} = 9.63$, the period calculated by Elton and Nicholson by considering the lynx population over the whole of Canada. Campbell and Walker (1977) go one step further, incorporating a second order autoregression as well as the estimation of the period ω^{-1} into their analysis, although their estimated period is also 9.63 years. In each case, the models chosen have significantly reduced the error sums of squares, although their forecasting abilities have not been demonstrated. Indeed, the "mixed-spectrum" model may not be appropriate since the second order autoregressive component of Campbell and Walker has a characteristic polynomial which has zeros whose arguments are $2\pi/18.5344$, corresponding to nearly twice the observed period of the data, so that the autoregressive component might be interpreted as a correction to the inadequacy of the sinusoidal component, rather than as a plausible explanation of the data.

Tong (1977) has considered the fitting of higher order autoregressive models to the data. Using Akaike's minimum AIC procedure he has obtained an eleventh order autoregressive model which fits the data quite well. Nevertheless, although Tong has considered also the best five-parameter subset autoregressive fit to the data, the large number of parameters would be difficult to interpret from a biological standpoint, especially since a lynx is considered to be "old" at ten years, and the lynx born in any one year are subject to an estimated annual mortality rate of 60%.

As noted in §8.1, the opinion of several authors is that a non-linear model might provide a better explanation of the data. In view of the models which have been considered, therefore, we have fitted a second order autoregressive model with random coefficients, that is

$$(8.2.3) \qquad X'(t) = (\beta_1+B_1(t))X'(t-1) + (\beta_2+B_2(t))X'(t-2) + \varepsilon(t)$$

where $X'(t) = X(t) - \mu$. The least squares estimates of μ, β_1, β_2, $\Sigma_{22} = E\{B_1^2(t)\}$, $\Sigma_{12} = E\{B_1(t)B_2(t)\}$, $\Sigma_{11} = E\{B_2^2(t)\}$ and $\sigma^2 = E\{\varepsilon^2(t)\}$ are respectively 2.9036, 1.3844, -.7479, .0821, -.0694, .0770 and .0364, while the maximum likelihood estimates are respectively 2.9036, 1.4274, -.8073, .0839, -.0489, .0664 and .0300. The arguments of the zeros of the characteristic polynomial $(1-1.4272z+.8073z^2)$ are $\pm 2\pi/9.6235$, so that a process satisfying (8.2.3) would be expected to exhibit a cycle having period 9.623 years, in close agreement with the estimate of Elton and Nicholson.

It is difficult to compare the previously considered models with the random coefficient autoregressive model, since $B_1(t)$, $B_2(t)$ and $\varepsilon(t)$ are not estimable separately. Moreover, since the best one-step-ahead predictor in the least squares sense for such a model is linear, the comparison of residual sums of squares might not be meaningful, because a linear least squares predictor will give a smaller residual sum of squares than any other linear predictor. The method of comparison adopted here is to estimate the various models using only the first one hundred observations, and then to compare the resulting one-step-ahead predictors with the realised values at the last fourteen time points.

The Moran-type model has been fitted using least squares for convenience, while the higher order autoregressive model which Tong has derived is a twelfth order model chosen as the best five-parameter subset autoregressive fit to the first one hundred observations. It should be noted that Moran has actually computed the prediction errors for the last eight observations, but has done so using the estimates obtained for the complete data set. The models of Moran and Tong are given respectively by (8.2.4) and (8.2.5):

$$(8.2.4) \qquad X'(t) = 1.3780X'(t-1) - .7486X'(t-2) + \varepsilon(t) ,$$

where $X'(t) = X(t) - 2.8802$ with $\hat{\sigma}^2 = .0572$, where $\hat{\sigma}^2$ is the estimated variance of $\varepsilon(t)$, and

$$(8.2.5) \qquad X'(t) = 1.0569X'(t-1) - .3374X'(t-2) - .0945X'(t-4) + .1512X'(t-9) - .1759X'(t-12) + \varepsilon(t) ,$$

with $\hat{\sigma}^2 = .0503$. The random coefficient autoregressive model is given by

$$(8.2.6) \qquad X'(t) = (1.4132+B_1(t))X'(t-1) + (-.7942+B_2(t))X'(t-2) + \varepsilon(t) ,$$

where $E\{[B_1(t)B_2(t)]'[B_1(t)B_2(t)]\}$ and σ^2 are estimated by $\begin{bmatrix} .0701 & -.0406 \\ -.0406 & .0492 \end{bmatrix}$ and .0391 respectively.

Tables 8.1 and 8.2 contain the results of the comparison between the five predictors considered, for the transformed and untransformed data respectively, while figures 8.2 and 8.3 represent graphically the data of table 8.1. In these tables and figures, the label MORAN 1 refers to the predictor obtained from (8.2.4):

$$2.8802 + 1.3780(X(t-1) - 2.8802) - .7486(X(t-2) - 2.8802) ,$$

while MORAN 2 refers to the non-linear predictor obtained in the manner indicated in (8.1.1) with $\gamma = 0$. N - Q1 denotes the linear predictor, obtained from the second-order random coefficient model (8.2.6):

$$2.8802 + 1.4132(X(t-1) - 2.8802) - .7942(X(t-2) - 2.8802) ,$$

while N - Q2 denotes the non-linear predictor $\tilde{X}(t) + 2.8802$, where $\tilde{X}(t)$ is the predictor of $X'(t)$ defined by (8.1.1). From these results we see, for the particular models compared here, that the non-linear predictors (MORAN 2 and N - Q2) for the transformed and untransformed data are better in the sense that the error sum of squares is smaller. For the two non-linear predictors, the N - Q2 predictor reduces the error sum of squares of the MORAN 2 predictor by five percent, with respect to both the transformed and untransformed data.

The comparison of the various predictors with the actual data has shown the advantages, in this case, of using non-linear models.

TABLE 8.1

ONE-STEP-AHEAD PREDICTORS OF THE TRANSFORMED LYNX DATA

YEAR	LOG_{10} LYNX DATA	MORAN 1	MORAN 2	TONG	N-Q1	N-Q2
1921	2.3598	2.4448	2.3835	2.4559	2.4596	2.3842
1922	2.6010	2.7971	2.6271	2.8088	2.8173	2.6323
1923	3.0538	2.8850	3.1193	2.8991	2.8989	3.0955
1924	3.3860	3.3285	3.3883	3.2306	3.3474	3.3971
1925	3.5532	3.4471	3.4955	3.3879	3.4571	3.4999
1926	3.4676	3.4289	3.4787	3.3321	3.4296	3.4781
1927	3.1867	3.1859	3.2683	3.0060	3.1759	3.2555
1928	2.7235	2.8628	2.6405	2.6875	2.8468	2.6587
1929	2.6857	2.4348	2.3747	2.4286	2.4153	2.3650
1930	2.8209	2.7296	2.5977	2.7643	2.7299	2.6292
1931	3.0000	2.9440	3.1277	2.9838	2.9508	3.0927
1932	3.2014	3.0897	3.1981	3.2169	3.0966	3.1762
1933	3.4244	3.2331	3.3065	3.3656	3.2390	3.2956
1934	3.5309	3.3896	3.4430	3.5035	3.3942	3.4413
Error Sum of Squares E		.2531	.2070	.2541	.2561	.1887
	$(E/14)^{\frac{1}{2}}$	.1344	.1216	.1347	.1353	.1161

TABLE 8.2

ONE-STEP-AHEAD PREDICTORS OF THE UNTRANSFORMED LYNX DATA

YEAR	LYNX DATA	MORAN 1	MORAN 2	TONG	N-Q1	N-Q2
1921	229	278	242	286	288	242
1922	399	627	424	644	656	429
1923	1132	767	1316	793	792	1246
1924	2432	2131	2445	1701	2225	2495
1925	3574	2800	3130	2443	2865	3162
1926	2935	2685	3011	2148	2689	3007
1927	1537	1534	1855	1014	1499	1801
1928	529	729	437	487	703	456
1929	485	272	237	268	260	232
1930	662	537	396	581	537	426
1931	1000	879	1342	963	893	1238
1932	1590	1229	1578	1648	1249	1500
1933	2657	1710	2025	2321	1733	1975
1934	3396	2452	2773	3188	2479	2762
Error Sums of Squares E		3029297	1384240	3100922	2711541	1319516
$(E/14)^{\frac{1}{2}}$		465.16	314.44	470.63	440.09	307.00

FIGURE 8.1
LOGARITHMS (TO THE BASE 10) OF THE LYNX DATA

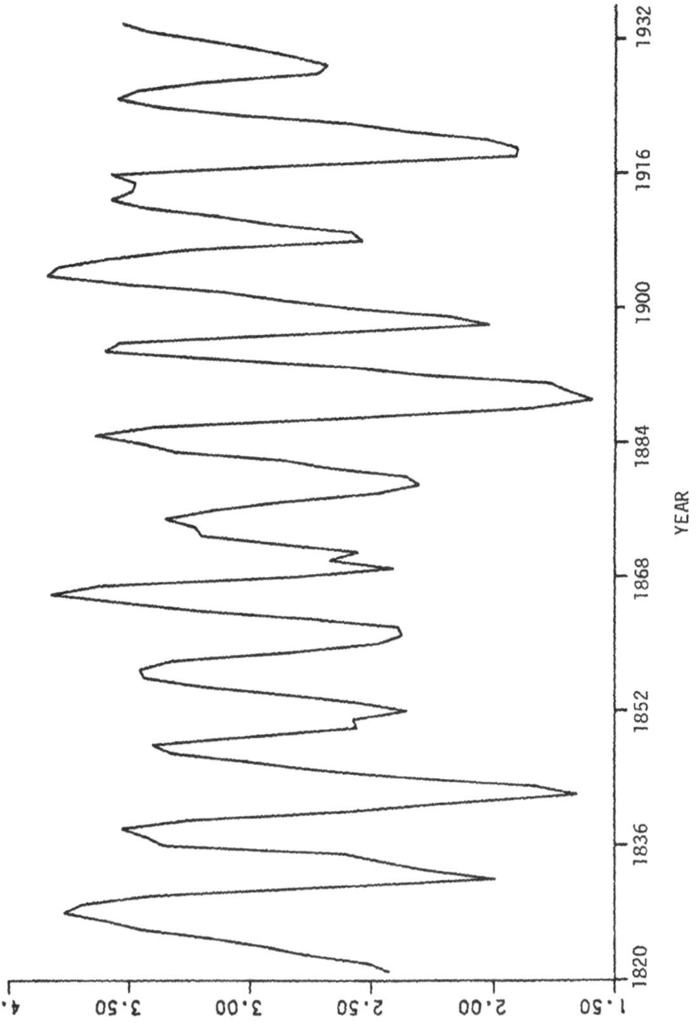

FIGURE 8.2

LINEAR AND NON-LINEAR PREDICTORS OF THE TRANSFORMED LYNX DATA

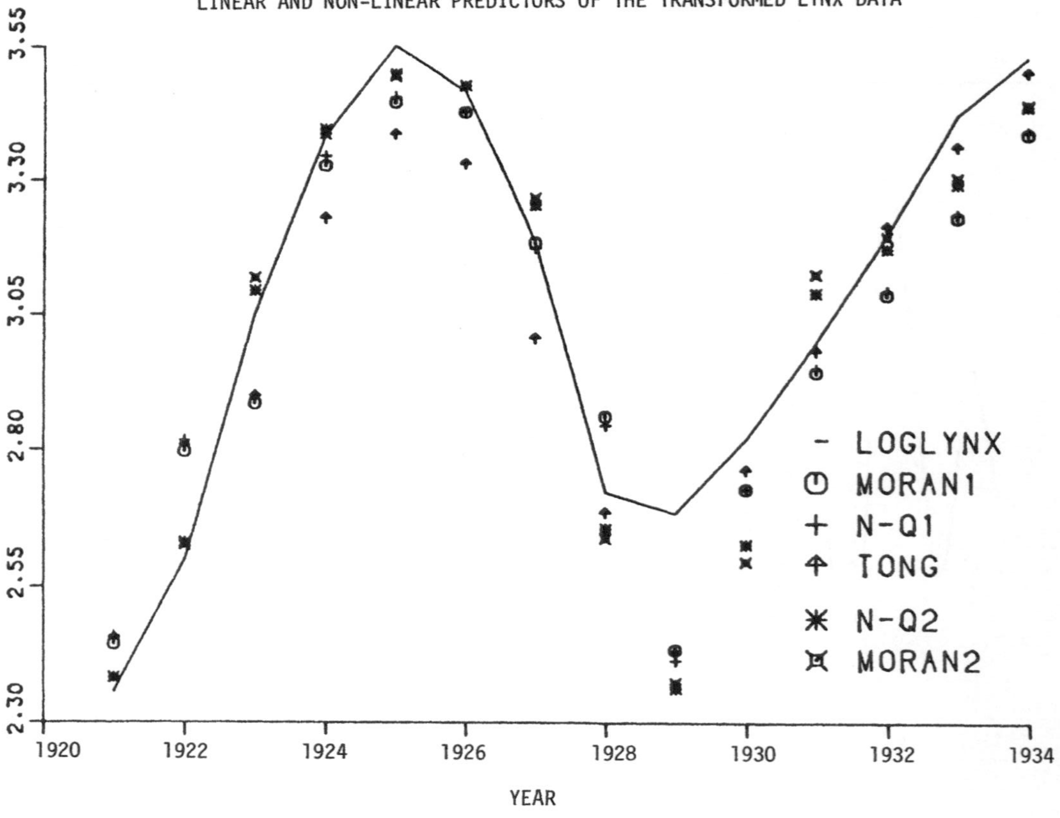

FIGURE 8.3

NON-LINEAR PREDICTORS OF TRANSFORMED LYNX DATA

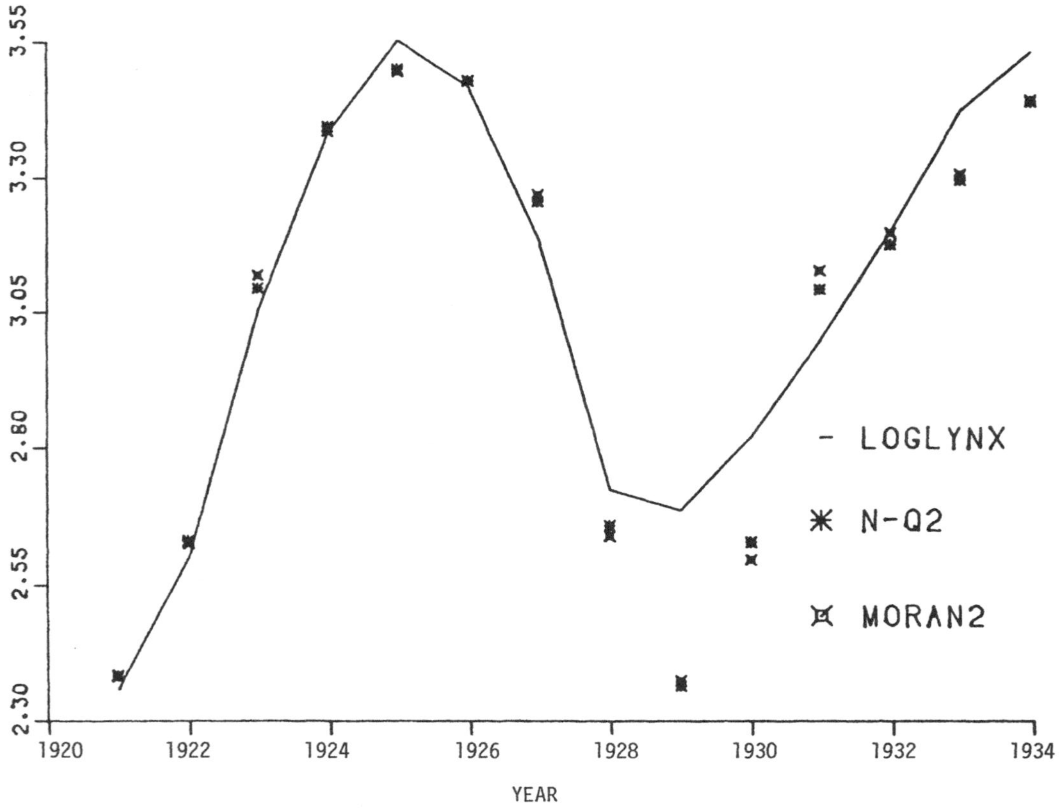

REFERENCES

Akaike, H. (1978), A Bayesian analysis of the minimum AIC procedure, *Ann. Inst. Statist. Math.*, 30, 9-14.

Andel, J. (1971), On the multiple autoregressive series, *Ann. Math. Statist.*, 42, 755-759.

Andel, J. (1976), Autoregressive series with random parameters, *Math. Operationsforsch. u. Statist.*, 7, 735-741.

Billingsley, P. (1961), The Lindeberg-Lèvy theorem for martingales, *Proc. Amer. Math. Soc.*, 12, 788-792.

Bulmer, M.G. (1974), A statistical analysis of the 10-year cycle in Canada, *J. Anim. Ecol.*, 43, 701-715.

Bulmer, M.G. (1975), Phase relations in the 10-year cycle, *J. Anim. Ecol.*, 44, 609-621.

Campbell, M.J. and Walker, A.M. (1977), A survey of statistical work on the Mackenzie River series of annual Canadian lynx trappings for the years 1921-1934 and a new analysis, *J. Roy. Statist. Soc.* A, 140, 411-431.

Chant, D. (1974), On asymptotic tests of composite hypotheses in non-standard conditions, *Biometrika*, 61, 291-298.

Conlisk, J. (1974), Stability in a random coefficient model, *Int. Econ. Rev.*, XV, 529-533.

Conlisk, J. (1976), A further note on stability in a random coefficient model, *Int. Econ. Rev.*, XVII, 759-764.

Elton, C. and Nicholson, M. (1942), The ten year cycle in numbers of lynx in Canada, *J. Anim. Ecol.*, 11, 215-244.

Finerty, J.P. (1980), *The Population Ecology of Cycles in Small Mammals*. Yale Univ. Press, New Haven.

Garbade, K. (1977), Two methods for examining the stability of regression coefficients, *J. Amer. Statist. Assoc.*, 72, 54-63.

Granger, C.W.J. and Hatanaka, M. (1964), *Spectral Analysis of Economic Time Series*. Princeton Univ. Press, Princeton.

Granger, C.W.J. and Andersen, A.P. (1978), *An Introduction to Bilinear Time Series Models*. Vandenhoeck and Ruprecht, Göttingen.

Hannan, E.J. (1960), *Time Series Analysis*. Methuen, London.

Hannan, E.J. (1970), *Multiple Time Series*, Wiley, New York.

Henderson, H.V. and Searle, S.R. (1979), Vec and vech operators for matrices with some uses in Jacobian and multivariate statistics, *Canad. J. Statist.*, 7, 65-81.

Jennrich, R.I. (1969), Asymptotic properties of non-linear least squares estimators, *Ann. Math. Statist.*, 40, 633-643.

Jones, D.A. (1978), Non-linear autoregressive processes, *Proc. Roy. Soc. London* A, 360, 71-95.

Kendall, M.G. (1953), The analysis of economic time series - Part I: Prices, *J. Roy. Statist. Soc.* A, 11-25.

Ledolter, J. (1980), Recursive estimation and adaptive forecasting in ARIMA models with time varying coefficients. In *Second Applied Time Series Symposium* (edited by D.F. Findley). Academic Press, New York.

Mehta, M.L. (1967), *Random Matrices and the Statistical Theory of Energy Levels*. Academic Press, New York.

Moran, P.A.P. (1953), The statistical analysis of the Canadian lynx cycle, *Aust. J. Zool.*, 1, 163-173; 291-298.

Moran, P.A.P. (1971), Maximum likelihood estimation in non standard conditions, *Proc. Camb. Phil. Soc.*, 70, 441-445.

Neudecker, H. (1969), Some theorems on matrix differentiation with special reference to Kronecker matrix products, *J. Amer. Statist. Assoc.*, 64, 953-963.

Neyman, J. (1959), Optimal asymptotic tests for composite statistical hypotheses. In *Probability and Statistics* (edited by U. Grenander), 213-234. Wiley, New York.

Ozaki, T. (1980), Non-linear time series models for non-linear vibrations, *J. Appl. Prob.*, 17, 84-93.

Pagan, A.R. (1980), Some identification and estimation results for regression models with stochastically varying parameters, *J. of Econometrics*, 13, 341-363.

Priestley, M.B. (1980), State-dependent models: A general approach to non-linear time series analysis, *J. of Time Ser. Anal.*, 1, 47-72.

Quenouille, M.H. (1947), A large sample test for the goodness of fit of autoregressive schemes, *J. Roy. Statist. Soc.* A, 110, 123-129.

Richter, H. (1958), Zur abschatzung von matrizennormen, *Math. Nachr.*, 18, 178-187.

Robinson, P.M. (1977), The estimation of a non-linear moving average model, *Stoch. Proc. and their Applic.*, 5, 81-90.

Rosenberg, B. (1973), A survey of stochastic parameter regression, *Ann. Econ. and Soc. Meas.*, 2, 381-398.

Subba Rao, T. (1970), The fitting of non stationary time series models with time dependent parameters, *J. Roy. Statist. Soc.* B, 32, 312-322.

Subba Rao, T. (1981), On the theory of bilinear time series models, *J. Roy. Statist. Soc.* B, 43, 244-255.

Tong, H. (1977), Some comments on the Canadian lynx data, *J. Roy. Statist. Soc.* A, 140, 432-436.

Tong, H. (1978), On a threshold model. In *Pattern Recognition and Signal Processing* (edited by C.M. Chen). Sijthoff and Noordhoff Int. Pub., The Netherlands.

Tong, H. and Lim, K.S. (1980), Threshold autoregression, limit cycles and cyclical data, *J. Roy. Statist. Soc.* B, 42, 245-293.

Turnovsky, S.J. (1968), Stochastic stability of short run market equilibrium under variations in supply, *Quar. J. of Econ.*, LXXXII, 666-668.

Author and Subject Index